THE SUPERNATURAL AND
ENGLISH FICTION

THE
SUPERNATURAL
AND
ENGLISH FICTION

Glen Cavaliero

Oxford   New York
OXFORD UNIVERSITY PRESS
1995

Oxford University Press, Walton Street, Oxford OX2 6DP
Oxford New York
Athens Auckland Bangkok Bombay
Calcutta Cape Town Dar es Salaam Delhi
Florence Hong Kong Istanbul Karachi
Kuala Lumpur Madras Madrid Melbourne
Mexico City Nairobi Paris Singapore
Taipei Tokyo Toronto
and associated companies in
Berlin Ibadan

Oxford is a trade mark of Oxford University Press

British Library Cataloguing in Publication Data
Data available

Library of Congress Cataloging in Publication Data
Cavaliero, Glen, 1927–
The supernatural and English fiction / Glen Cavaliero.
  p. cm.
Includes bibliographical references and index.
1. English fiction—History and criticism. 2. Supernatural in
literature. 3. Fantastic fiction, English—History and criticism.
4. Horror tales, English—History and criticism. 5. Gothic revival
(Literature)—Great Britain. I. Title.
PR830.S85C38 1995 823.009'37—dc20 94–5272
ISBN 0–19–212607–5

10 9 8 7 6 5 4 3 2 1

Typeset by Graphicraft Typesetters Ltd., Hong Kong
Printed in Great Britain
on acid-free paper by
Bookcraft (Bath) Ltd.
Midsomer Norton
Avon

for Anne Lee
and in memory of John

PREFACE

This book derives from an observation and a query. The observation notes the repeated tendency of English novelists to write about the supernatural, or at any rate about mysterious and inexplicable events. Being a multi-racial people with contradictory cultural impulsions, and living in an erratic climate, we produce a literature of ambiguities and borderlines. Dilapidated houses, forsaken upland valleys, far-off wailing heard at night upon a lonesome sea-shore, transcendental ecstasies in summer meadows, the furtive shifting of a headstone in a city graveyard —these are among the metaphors resorted to by those writers who seem unable to restrict their view of life to the confines of outward appearances and rational deduction. For all the perennial popularity of such steady-eyed observers of social forces and relationships as Trollope and Jane Austen, English novelists are haunted by the presence of mystery and strangeness; they even, as in the cases of Dickens and James Joyce, render the surface plausibilities of naturalism with an artifice of such elaboration as to suggest impatience with the mean scrupulosity involved in the preservation of conventional appearances. From the days of Horace Walpole to those of Iris Murdoch the urge towards celebrating the other-worldly and abnormal is unceasing. Such themes are not eccentric to English fiction but are among its recurring and characteristic features.

As to the query: is this attraction towards what one may loosely call the supernatural an attraction towards a comprehensive understanding of the real, or does it amount to a delusion? Is it an aberrant irrelevance? or an instinctive witness to a metaphysic for which the presuppositions of a literal-minded materialism make no allowance? Such questions involve the status of fiction itself as a trustworthy medium of communication. The history of the supernaturalist novel throws a good deal of light upon the concept of fictive truth and upon the validity of fictional representations.

The emphasis in this book rests on the impact of metaphysical themes and subject-matter on naturalistic fiction. Fantasies are disqualified from discussion, even though some of them contain supernaturalist elements, since, by dealing in avowedly imaginary or eccentric worlds, they pre-empt the particular relationship between truth and fiction which is my concern; nor have I included tales of the magical or marvellous, such as William Beckford's *Vathek* (1786), Rider Haggard's *She* (1887), or the stories of H. G. Wells, which are so many steps on the road to the entirely self-sufficient categories of space and science fiction.

Books for children present a problem. Although they frequently portray the presence of the supernatural in a context more persuasive than the adult novel can provide, that very persuasiveness reduces the complexity of the interaction between the two categories of the naturalistic and the metaphysical. In E. Nesbit's tales, for instance, the various preternatural beings—the Psammead, the Mouldiwarp, and the rest—are as humorously personified as are the human characters. The few novelists for children whom I have discussed are those whose work suggests a genuine metaphysical reality operative within a naturalistically presented social world. In addition I have included a number of writers, Phyllis Paul most notably, whose novels are not as well known as they deserve to be.

Books of literary criticism are often advertised as being intended 'both for academics and for the general reader'—a distinction it takes an academic mind to draw. While I have tried to effect a discreet deployment of the particular kind of language which, for purposes of concision and lucidity, students of literature have evolved for use among themselves, I hope that any reader who enjoys novels as much as I do, and who takes them with a corresponding seriousness, will find this handling of the theme neither obscure nor superficial. I am conscious that I have not discussed these writers with the comprehensiveness that they deserve; but at a time of specializations like the present I am inclined to believe that a general survey of this particular terrain could prove to be of use.

With regard to the book's design, the opening chapter enlarges on some of the issues raised above, discussing the concept of naturalism ('truth to life') and the relevance of supernaturalist fiction to contemporary literary theory and its ideological accompaniments; it then proceeds to make certain definitions and distinctions as to the terminology of such aspects of the supernatural as 'mystery', 'the preternatural', and 'the paranormal'. Most previous critical work on the subject has been faced with the difficulty of determining on a suitable vocabulary, and understandably so, for this is an area where philosophy, theology, and metaphysics meet. Some people may find this section unpalatable or even meaningless, since it employs a mode of discourse that is currently unfamiliar, albeit necessary if the concept of supernaturalism is to be discussed in terminology appropriate to it. I have treated that concept on its own terms, and have not been concerned to criticize or deconstruct it, but rather to observe its operations within the imaginative writings of English novelists of all outlooks and persuasions.

In the ensuing chapters the method followed is a fusion of the historical with the descriptive and the analytic. The reasons for including the various writings under their several ascriptions should become apparent as the argument proceeds. To begin with, there is an account of those novelists, mainly from the nineteenth century who, working from within the Gothic tradition, treated the supernatural as something menacing and disruptive, as, in short, the preternatural. They show little or no concern with metaphysical reality as such. Theirs is the world whose imaginative expression is mediated through the imagery of ruined abbeys, embattled castles, haunted houses, creatures of nightmare—a world of whispers and rustling, of cramped spaces, dark passages, and wild weather in treacherous and forsaken landscapes. It is perhaps what most people think of first when they hear the word 'supernatural'.

Chapter 3 focuses on the hermetic tradition, one which treats metaphysical reality with systematic seriousness, mining that huge cultural underground of mystical traditions, and of those societies which adopt their teachings and spiritual secrets. Its crudest imaginative expression is in portrayals of magic, witchcraft, and spiritual manipulation; more profoundly, it reflects the world of syncretist symbolism that draws on the cabbala (the occult lore of Judaism), on Rosicrucianism and the mystical Grail quest, on theosophy and spiritualism. This is the world of clairvoyance, of seances and second sight, all of which are under levy by authors in the hermetic tradition in order to provide material for a depiction of the supernatural as humanity's spiritual homeland, with a hidden body of laws that are operative throughout the universe.

The succeeding chapter deals with the adoption of super-naturalist themes by a number of novelists in whose work these elements are integrated into the scheme of traditional theology, both Protestant and Catholic. Their inbuilt intellectual accept-ance of metaphysical reality leads them to exploit supernaturalist themes for purposes of fable: what matters to them is not meta-physics as such, but spiritual morality. In this respect they differ from the hermetic school, their fictions being exemplary rather than exploratory in character.

Chapter 5 enters a world of imaginative borderlands, of experi-mentation. Two writers are discussed here: Rudyard Kipling, whose short stories cover an extremely wide range of supernatu-ral manifestations in a variety of moods and in the context of a thoroughly contemporary naturalism, and whose attitude towards metaphysical reality is inquisitive, observant, and responsive; and Walter de la Mare, whose work portrays with well-nigh obsessive concentration the ambiguous nature of the twentieth-century Western consciousness, haunted as it is by spiritual forces it is unable either fully to understand or to accept. Well described by George Steiner as a 'master of whispering', he has an attitude towards metaphysics that is both restive and bereaved.

The three chapters which follow deal with novelists who pro-vide new contexts for supernaturalist writing. Those discussed in Chapter 6 respond in various ways to the suggestive power of landscape: theirs is the world of geomancy, earth-worship, fertil-ity cults, local legends, and pagan survivals of all kinds—also of romantic epiphanies in response to natural beauty. They present metaphysical reality not as transcending the material world but as immanent within it. There follows an account of a number of writers who deal in the world of consciousness, of delusion and illusion, of the workings of guilt and psychological distur-bance; these are introverted authors as compared with the more outward-looking neo-pagan ones. For them, metaphysical realities are internalized; theirs is more of an indoor than an outdoor world, one that has become engulfed in the supernatural dimension.

The chapter called 'God-Games' discusses the deliberate use of supernaturalist material in order to construct theories as to fictive truthfulness by novelists who exhibit linguistic relativities and a self-conscious use of allegory and fable. This is the current cer-ebral world of late twentieth-century agnosticism, the world which implicitly rejects any sense of metaphysical reality. None the less, some of the most distinguished exemplars of the genre—William Golding, Muriel Spark, Iris Murdoch—are notable for an acceptance of Christianity or a sympathetic response to it. This is a chapter of unfinished business, despite which the following one traces the continuation of the various supernaturalist literary traditions into the present time. (I use 'supernaturalist' in the way that one speaks of a 'realist' as distinct from a 'realistic' novel: the term refers to genre rather than to methodology.) A final chapter examines the structuring of supernatural experience as it is presented in fictional terms, and assesses the value of such inclusiveness for the naturalistic novel as a literary form.

Two final points. The influence on English writers of the American supernaturalist tradition—Charles Brockden Brown, Hawthorne, Poe, Ambrose Bierce, and H. P. Lovecraft, to name its most celebrated exemplars—is only mentioned parenthetically: it would require a whole book to itself. As to the term 'English', objections may be raised to its use in the title. I had thought of writing 'British'; but that designation seems so irrevocably associated with matters social and political that it seemed more sensible to retain the familiar one as being, however loosely, appropriate to a work of fiction written in the British Isles—in English.

In conclusion I must record my particular thanks to Dr Paul Hartle and Dr Helen Small, both of whom read this book in manuscript and made many helpful suggestions towards its improvement. I am also indebted for various kinds of help to Dr Marius Buning, Mr Roderick Cavaliero, Miss Esmé Cogdon, Mr Michael Cox, Mr Mark Gibson, Mr Peter Gregory-Jones, Mrs Mildred Hartridge, Mr Brian Jenkins, Miss Lois Lang-Sims, Professor Charles Lock, Mr Jim McCue, Mr Peter Scupham, Dr John Shakeshaft, and the Reverend Ian Tattum.

CONTENTS

The Whole Creation is a Mystery, and particularly that of
Man.

(Sir Thomas Browne)

Ah, what a dusty answer gets the soul
When hot for certainties in this our life!
(George Meredith)

Man can embody truth, but he cannot know it.
(W. B. Yeats)

Now faith is the substance of things hoped for, the evidence
of things not seen.
(Epistle to the Hebrews)

1

THE JOKER IN THE PACK

'Well, it's no use your talking about waking him,' said
Tweedledum, 'when you're only one of the things in his dream.'
(Lewis Carroll, *Through the Looking Glass*)

Natural and Supernatural: Wuthering Heights

'All true histories contain instruction.' The opening words of Anne
Brontë's *Agnes Grey* differ markedly from those of the more
famous novels of her sisters. Her voice is judicious, detached; in
Jane Eyre and *Wuthering Heights*, on the other hand, the call on
one's attention is peremptory. 'There was no possibility of taking
a walk that day', and '1801:—I have just returned from a visit to
my landlord' are sentences that immediately immerse one in the
narrator's preoccupations; and it is this sense of close, solicited
involvement, of being in a physically real and not a fictive world,
which is a distinguishing mark of what may be called the Brontë
spell. Even the opening of *The Tenant of Wildfell Hall* conveys it—
'You must go back with me to the autumn of 1827'—but what
follows in that novel, as in *Agnes Grey*, in its sober, controlled
delineation of character in a naturalistically observed location,
distinctively belongs to Anne—not only the experience of her
brother's self-destructive drinking, but also her response to it,
the calm, steady-eyed adherence to duty which was part of her
Calvinistic upbringing, and which denied her any fictitious consolations or embellishments. The world was in the hands of God:
truth was, therefore, instructive by necessity, and there was no
call to improve upon it, since all contingency was ultimately
providential. The flesh was, in that particular creed, subsumed
by spirit: there was no question of an intermediary plane of being.

Life as a whole was supernatural, not only the miraculous or inexplicable events within it.

There is therefore a certain irony in the fact that *Agnes Grey* should have first appeared in tandem with a novel in which uncanny and mysterious happenings play a vital part. *Wuthering Heights*, however, treats them not as something intrusive or abnormal but as an integral aspect of a realistically presented social world. The conjunction of the two novels is thus appropriate as well as piquant. Emily Brontë's book, while appearing to contrast with the straightforward narrative of her younger sister, is, for all its popular reputation as a romantic melodrama, a signal instance of the fictive naturalism to which Anne was more obviously faithful. The violent behaviour of its characters is so realistically portrayed that it startled the author's contemporaries as much as did the unorthodoxies in *The Tenant of Wildfell Hall*; but unlike the latter book, *Wuthering Heights* penetrates behind appearances to propound the existence of a spiritual reality, one more objective than could be provided by a simple assurance of personal salvation or by viewing the history of the physical universe as the work of the controlling hand of Providence. The book's complexities and seeming contradictions arise from its depiction of the familiar material world not as being distinct from spirit but as being porous to it. It is this fusion of surface naturalism with a metaphysical mode of discourse which makes the novel the unique thing that it is; but it is the naturalism which validates the metaphysical assertions.

To claim that *Wuthering Heights* is a supernaturalist novel may seem wrong-headed and contentious; and the book certainly contains sufficient elements of obviously authentic detail as almost to warrant its being read primarily as a tale of late eighteenth-century Yorkshire rural life, with its probable origins in a real-life domestic drama at Walterclough Hall near Halifax.[1] As a naturalistic novel it satisfies through its apparent spontaneity and informed use of local speech and manners; but it is also naturalistic about certain subjects—religion, family life, the reverence for femininity—which contemporary literary convention tended to regard as sacred and only to be handled in the mealy-mouthed manner encouraged by the circulating libraries. Emily Brontë does not decry them; but her concern is with what lies behind them. *Wuthering Heights* supplies a critique of idolatrous materialistic values, but nowhere in unqualified or simplistic terms.

A first glance at the book suggests that it is a localizing, even a domestication, of those Gothic tales of folklore and wonder to be found in *Blackwood's Magazine*, which used to be passed on to the Reverend Patrick Brontë by a local doctor. But even though aspects of Lockwood's night in the 'haunted' chamber recall such stories, one notes that the chamber in question is in fact merely of the simple farmhouse kind with a 'press' bed in it, and that in any case all the conventional romantic aspects and trappings are the product of the equally conventional Lockwood's imagination. The author uses these elements familiarly, so as to make clear her view of the preternatural as being, to quote David Cecil, 'a natural feature of the world as she sees it',[2] and as such of no extra metaphysical significance. Lockwood's dream, although eerie through its very homeliness and particularity, is earthed by the previous farcical yet unnerving nightmare about 'the famous Jabez Branderham' and his sermon in 490 parts (which may reflect the author's disrespectful attitude to the Calvinistic Methodism of her Aunt Branwell). This dream is accounted for by the rattle of fir-cones against the window; but that explanation is reversed when the slumbering Lockwood, in order to silence it, breaks the window and is clasped by the ghostly hand of Catherine. The episode grows more extraordinary as one realizes both the disturbing fact that the dream has turned out to be true, and also the outrage perpetrated by the angry Lockwood upon Heathcliff's susceptibilities; it is as though the potentially melodramatic intensities of the tragic story are being subjected to a pre-emptive serio-comical subversion.

Later in the novel the supernatural is encountered more directly still in an aspect which stresses its normality in terms of the psychic worlds of the protagonists. Cathy's spirit joins Heathcliff as he wrenches at her coffin lid: 'I felt her by me—I could *almost* see her, and yet I *could not*! . . . She showed herself as she often was in life, a devil to me!' (ch. 29). This kind of close relationship is something far more real, more quasi-physical, than the usual apparitions of the dead; it suggests an interaction of personalities, and is more akin to Hardy's poem, 'After a Journey',[3] than it is to the visitation to Milton of his 'late espoused Saint',[4] and reaches its consummation in Heathcliff's final communing with Catherine before his death.

In *Wuthering Heights* preternatural elements coalesce with visionary ones. Whether Lockwood physically hears the ghost of

Cathy crying at the window, or whether he only dreams about it, is secondary to the impression that in supernatural reality her spirit is there, as in her dying ravings she had predicted that it would be. And even more powerful than the presence of the haunter is the anguish of the haunted. Where Cathy looks in at the closet window, so Heathcliff gazes out: something unseen mediates the reality in which they share. A similar effect is achieved when the village boy sees the lovers' ghosts upon the moor. He may only be reflecting superstitious gossip; on the other hand, the sheep refuse to pass the spot. Both cases suggest a spiritual presence within the physical dimension, and both refuse any simplifying resort to physical verification such as would render these events departmental or superfluous. Nothing in the novel is merely sensational or spooky; and the supernatural happenings which seem to colour so much of the narrative are in fact singularly few—so few as almost to justify the contention of one critic that what she calls 'the "metaphysical" parts' of the book are 'brief, and on the whole misleading'.[5]

It is more by its structure and technique that *Wuthering Heights* conveys the sense of a supernatural dimension. Its use of retrospect and interpolated narrative, the variety of points of view (Lockwood's, Nelly Dean's, Isabella's, Catherine Linton's) alike imply the relativity of chronological time and of the existence of some eternal present in which earthly events take place. The avoidance of sequential chronology and the circular movement of the action, reinforce this effect. Likewise, the repetition of certain symbolic motifs suggests that things happen simultaneously on different planes. The discovery that Heathcliff on his deathbed has grazed his hand on the loose window-frame, recalls Lockwood's encounter in the same place with the ghost of Catherine; in lacerating her wrists he is enacting Heathcliff's own life-in-death agony.

But if *Wuthering Heights* portrays an experience of a specifically individual kind, it also envisages a more general supernatural reality, beneficent and providential, which becomes apparent in the prosaic Nelly's vision of the face of Hindley Earnshaw by the guide stone on the moor, when old affections and associations quicken her sensibility, and promote her return to the Heights. An overriding mystery encompasses all the differing forms of supernatural experience which the book contains.

As to the famous concluding paragraph, it is a classic instance of a statement governed and modified by the tone of its delivery.

I lingered round them, under that benign sky; watched the moths fluttering among the heath and hare-bells, listened to the soft wind breathing through the grass, and wondered how any one could ever imagine unquiet slumbers for the sleepers in that quiet earth.

The lulling cadence determines the tranquillity of the ending; yet all it tells us specifically is that the narrator himself cannot imagine any preternatural appearances being associated with such a place. But the cadence contradicts the notion that the language purports to convey, and suggests an affirmative rather than a negative, albeit resigned, conclusion. The whole novel works towards the realization that what Heathcliff and Catherine are in death, they have been, or are, in life: death is but the true condition of, or verdict upon, those lives.

What sets *Wuthering Heights* far above the average romance of tormented loves and passions and supernatural doom, however, is its humanity. For if Heathcliff operates as a demonic force, the language he uses is that of an ordinary suffering man.

I have neither fear nor a presentiment, nor a hope of death. Why should I? With my hard constitution and temperate mode of living, and unperilous occupations, I ought to, and probably *shall*, remain above ground till there is scarcely a black hair on my head. And yet I cannot continue in this condition! . . . O God! It is a long fight, I wish it were over. (ch. 33)

One observes the specificity, the colloquial ease, even in describing so extreme a state. When Nelly wonders if he be a ghoul or a vampire, she goes on to reflect that 'I had tended him in infancy, and watched him grow to youth, and followed him almost through his whole course; and what absurd nonsense it was to yield to that sense of horror' (ch. 34).

Wuthering Heights constantly evokes a sense of the weird and unearthly, only to reaffirm the wholesome sanity of ordinary experience. Thus the Calvinistic reading of Heathcliff as a damned soul is endorsed not in theological terms but in the constriction of his position and of his personal unhappiness; he is damned in a purely psychological, material sense. His ill-treatment by Joseph is a caustic satire on the materialistic self-righteous aspects of contemporary religion. But the demonic aspects of Heathcliff's character are subtly suggested, never affirmed. It is his unknown origin and the puzzle of his adoption by the elder Earnshaw which lend him the qualities of a changeling that others apply to him; it is his disappearance and return as a wealthy man that are reminiscent of a folk-tale. However dour and angrily impassioned, he is always fully human, even at times pitiable. His relations with Cathy resemble those of two perpetually ungovernable children. It is Isabella, not Catherine, who has the conventional romantic passion for Heathcliff, and it is she who sees him in demonic terms. In this she contrasts with Nelly, whose voice— limited, affectionate, matter-of-fact, and with a touch of York-shire astringency—sets the prevailing tone.

And yet even Nelly can be regarded by the delirious Catherine as a witch. 'So you seek elf-bolts to hurt us! Let me go, and I'll make her rue! I'll make her howl a recantation!' (ch. 12). This has a Shakespearean ring. By their frequent deployment as terms of casual abuse such epithets are deprived of their preternatural potency. Similarly, Cathy's cry, 'I *am* Heathcliff!' is more telling, and more persuasive as to the existence of a spiritual universe, than is the apparition which the materialistic Lockwood thinks he sees. All the apparitions are sacramental—they both are (or can be) what they appear to be, and they are also instrumental in conveying the reality to which they draw attention. In *Wuthering Heights* the spiritual world is the real world, portrayed with a conviction that is inescapable and not easily to be explained away. It is vital to the story. Chesterton's dictum applies here: 'Take away the supernatural and what remains is the unnatural.'[6]

And yet 'Only perceive purely and the spiritual and the material world vibrate as one.'[7] Such a conflation transfigures the ideal of faithful naturalism which the first readers of the book were to encounter when they proceeded to 'Volume Three' and the tranquil paragraphs of *Agnes Grey*. In that novel material appearances are spiritualized not by metaphysics but by conduct. When its narrator's story comes to a conventional end in marriage, she concludes with a profession of simple adequacy. 'And now I think I have said sufficient.' It is a model of civilized, rational, self-contained, good literary manners.

Wuthering Heights would seem to outrage those manners, and to refute the orderly foreclosures of conventional fiction: no wonder that Charlotte Brontë likened it to something 'terrible and goblin-like',[8] and it has been regarded as a magnificent oddity ever since. But it is one only in this respect, that it combines a metaphysical vision with a naturalistic technique: Emily Brontë's mysticism, like that of all genuine visionaries, is rooted in a full acceptance and command of the material world. Certainly we are at no time in an eccentric or specialized or a private universe, less so, indeed, than we are in a novel by Dickens, or even in one by such an admitted naturalist as Henry James.

However, *Wuthering Heights* is unique, and remains a measure by which all later supernaturalist novels can be assessed. In part this is because it takes the slapdash worldly notion of supernatural religion (life after death as the reward of good behaviour and the punishment of bad, with a few random diabolic agencies loosed to torment the disobedient on earth) and replaces it with the concept of natural spiritual affinities and an all-embracing spiritual universe. The supernatural encompasses the characters without intruding upon them or controlling them; and there are few if any other novels of which this can be said. Some possible examples will be discussed in due course; but in each instance the supernatural, to varying degrees of effectiveness and persuasiveness, either plays a part *within* the action or else dominates that action altogether. Whichever be the case, the reader is to some extent under compulsion either to accept or to reject it, since the plausibility of the text itself is bound up with such a choice.

As an element within naturalistic fiction, the supernatural has a teasing, confrontational quality that upsets the self-sufficiency of naturalistic art and the materialistic philosophies of which it is an expression. *Wuthering Heights* in particular disconcerts, because, as Heather Glen remarks, it contains 'a foregrounding of questions which in other mid-Victorian novels—indeed in much of our thinking today—are largely unconsidered'.[9] So should this novel be regarded as eccentric? Or is its perspective more truly naturalistic than a philosophy of simple rationalism will allow? Certainly it epitomizes most of the issues raised for English fiction by the supernaturalist tradition as a whole, and raises the question as to the true scope and authority of traditional materialistic naturalism as determining individual and social responses. Is any comprehensive portrait of human nature and society really possible? Or is the naturalistic novelist simply a conjuror among other people's conjurations? The questions facing literary criticism involve philosophy and conduct as well as, and as part of, the relation of language to ideology, and are never so purely literary as most critics, probably, would have them be.

Literary Theory and the Supernatural

The call for a work of fiction to be truthful might seem to be a paradox, were it not for the capacity of novels to influence behaviour as well as simply to reflect it. Anne Brontë's hope, in the words of her heroine, that her history 'might prove useful to some and entertaining to others' satisfies Samuel Johnson's requirement that a piece of writing should help one 'the better to enjoy life or the better to endure it'.[10] In this respect her art is an eighteenth-century art; but it eschews the pragmatic, the improvisatory, and the parodic in a most un-eighteenth-century way. In the early 1920s George Moore, self-appointed apostle of French naturalism, was to extol *Agnes Grey* as 'the most perfect prose narrative in English literature . . . style, characters and subject are in perfect keeping'.[11] Leaving aside Moore's notorious addiction to provocative assessments, such a contention reflects a naturalistic writer's ideal that, while novelists should portray in faithful detail the observable facts of physical reality, their own artistry should be invisible, so that no gap would intervene between their subject and its rendering, still less between that material and their personal emotions. There is a puritanical fineness of purpose in the aspiration. The aim of fiction was to convince the reader of its self-sufficient truthfulness; authorial control should be subservient to that single purpose. The novelist must act as a just and righteous God would act to His creation, with the simplicity of single-mindedness, self-expression bound up irrevocably with the welfare of the artefact.

But men are not gods, and as Moore goes on to remark, 'The simple is never commonplace.' Such requirements as to disciplined impersonality are rarely satisfied—or satisfying: *Agnes Grey* has never enjoyed more than a limited esteem, while Moore's own *Esther Waters*, although deservedly a landmark in the history of the English novel, would seem to be nobody's particular favourite. Neither book contains any trace of the mysterious or unpredictable. In this they are faithful to the naturalistic mainstream of fiction; and yet that tradition continues to break out of the strait-jacket of materialistic thought and feeling, a close attention to received appearances being an insufficient measure for what human life affords. For each George Moore one finds a Frederick Marryat who refuses to confine himself to plain tales of the sea, and who writes *The Phantom Ship* (1839), with its werewolf story, and *Snarleyow; or The Dog Fiend* (1837) (that serio-comic account of an obsession which can still induce a shiver of disquiet). Even the worldly-wise Anthony Trollope saw fit to provide a consolatory spiritual visitation for the grieving hero of *Marion Fay* (1882), while the confident expanse of Edwardian naturalism was gently ruffled by the obtrusion of *The Ghost* (1907) and *The Glimpse* (1909), novels which reflect the interest taken in the occult by the arch-champion of fictive realism, Arnold Bennett. Nor was George Moore himself immune from an interest in the spiritual forces motivating the religion which he did not share, witness *Sister Teresa* (1901) and *The Lake* (1905). As to Daniel Defoe, who is usually dubbed the founding father of the English naturalistic novel, he too instils his fiction with a sense of the unseen, whether through premonition, as in *Moll Flanders* (1722), through second sight, as in *Roxana* (1724), or through an emerging belief in a controlling providence, such as that which comes to the protagonist of *Colonel Jack* (1722), who at the completion of his story finds that 'in collecting the various Changes and Turns of my Affairs, I saw clearer than ever I had done before, how an invisible overruling Power, a Hand influenced from above, governs all our Actions of every Kind, limits all our Designs, and orders the Events of every thing relating to us.' To change the metaphor, he discovers the author of the book that is his life. It is the same sense of providential guidance as that which shapes the narrative of *Agnes Grey*.

Although so many distinguished writers have concerned themselves with the supernatural, criticism tends to regard metaphysical questions as ornamental or—more justifiably in view of the proliferation of a preternaturalist literary subculture—as departmental. The subject is a source of confusion, and is accordingly dismissed as eccentric or irrelevant. It has been a central concern of poetry from earliest times, some would say its source;[12] but because novels became generally popular at a period when Deism was secularizing the natural order, with scientific materialism filling the gap left by an absentee divinity, any supernaturalist element in a work of prose fiction is inclined to compromise that work's credentials (credentials based on truthfulness

as to perceived reality and to human behaviour); and the willing-
ness to accept a supernaturalist ingredient as itself constituting
part of that truthfulness inevitably diminishes as popular and
intellectual attitudes dictate.

Under the successive impacts of Marxist, structuralist, and
post-structuralist criticism the topic has been devalued to the
point of being discredited. Marxism views novelists primarily as
class-conditioned, stressing their importance not so much in terms
of what they have to say about the individual aspects of human
personality as in terms of their relationship with their social
environment: the approach is utilitarian and rooted in dogma.
Structuralists, on the other hand, designate novels 'texts', objects
of non-ideological study for their patterns of verbal relationships
and linguistic nuance; while post-structuralists treat them as
phenomena to be broken down into their constituent parts. These
include the social and psychological factors that go to their
making, and those which condition their reception, such as
reading-habits, critical expectations and academic fashions, mar-
keting and book production, all of which contribute to the verbal
complex summed up in the work 'novel' (and all of which are
equally applicable to the critic's own text and pre-emptive expectations). A book is currently not so much a structure as an event;
and in the process a commonly held body of assumptions with
regard to truthfulness and worth is lost. We are awash in a sea
of relativities.

The place of supernaturalistic fiction within the traditional novel
as a whole corresponds to the place of fiction itself within prose
writing as a whole. For the distinction between fictiveness and
truthfulness can itself be called in question: all absolutes in this
area of discourse would seem to be dissolved, both by the realization that language generates its meaning no less than it
conveys it, and by the demonstration of the relativity of texts
and of their varying conditions of production and assimilation.
Accordingly, in the light of those dissolutions, and of the accompanying queries as to the validity of traditional critical approaches
and vocabularies, the introduction of assumptions as to metaphysics becomes scarcely less permissible or less verifiable than
is the practice of a fictive naturalism.[13] Indeed, the acceptance of
such relativity may further, rather than impede, an understanding
of supernaturalism and of its function in the literary procedures
of the novel form as a whole.

To talk meaningfully about the supernatural calls for a basis of metaphysics; and that particular basis is usually either lacking or misunderstood, not to say disallowed for being irrelevant. Whatever may be the case in contemporary academic discourse, writers and readers alike do have their own assumptions as to spiritual reality and live their lives in accordance with them, often in the face of what they may outwardly profess. Novelists deal not only with social and personal relationships, but also with the individual consciousness alone with itself; and this consciousness operates in a different manner from that in which the critical intelligence deploys its various strategies and theories—a point which supernaturalist writers often make. As it is, where supernaturalist themes are concerned, contemporary responses tend to be blinded by science, as helpless in their reactions as an improperly fed computer. Such a comparison becomes still more appropriate when one considers that computers think consequentially and cannot reason laterally, and that it is lateral thinking that is at the root of philosophy and metaphysics, and of much artistic creativity. But, as Lafeu remarks in *All's Well That Ends Well*, 'We have our philosophical persons, to make modern and familiar, things supernatural and causeless' (II. iii. 1).

It follows that the treatment of supernaturalist themes in the predominantly naturalistic novel is relevant to an understanding of the nature of fictive truth, not only as it affects the art of the novel, but also as it impinges upon critical theory and upon the presuppositions of contemporary society. In isolation, literary theory can certainly become a form of atheistical scholasticism, turning in upon itself in self-perpetuating self-sufficiency; but at least it does not, as more overtly ideological procedures tend to do, signal through veils towards some limiting, and thus delusive, spiritual presence. Deconstructive literary methodologies can reduce linguistic formulations to mere 'constructs of supposition', which as such 'cannot transcend the medium of their own saying';[14] and the liberal humanist tradition, with its insistence on 'spiritual values' necessarily rejects such intellectual solipsism. But that rejection has a materialistic bias. It fetters the leap of faith (necessary if the concept of transcendence is to be definable at all) to an essentially idolatrous understanding of the term 'spirit', reifying it and thus subjecting it, as though it were an object, to the relativities of competing ideologies. Mystics of all schools know otherwise; and those supernaturalist novelists who

devote themselves to serious metaphysical enquiry reveal the
fluctuating nature both of human responses to the dimension of
mystery and of the terms on which its operations may be apprehended. As such, their writings form an affirmative counterpart
to the negative findings of deconstructive criticism. Their dramatizing of supernaturalist experience demonstrates that the roots
of metaphysical enquiry are found in diversified responses, and
that while no hypothetical intellectual proposition can constitute
the absolute, it may constitute *an* absolute that is relative to
other hypotheses. But all alike are relative to an absolute that by
definition is unknowable.

In contemporary linguistic philosophy the transcendent, when
personified as 'God', is dismissed for being a human self-
projection; instead it is intuited as negativity, as the emptiness or
no-thing which defines the substance of being, in all its manifold
forms. But to identify 'God' with that principle of separation is
itself idolatry, for the absolute transcends every human formulation, even that of nothingness. Accordingly, to quote Rudolf Otto,
'in neither the sublime nor the magical, effective as they are, has
art more than an indirect means of representing the numinous.
Of directer methods, our Western art has only two, and they are
in a noteworthy way *negative*, VIZ, *darkness* and *silence*.'[15] The
dark night of the soul may issue in an unimaginable lucidity; but
in terms of fiction and philosophy the relativities of verbal formulations, which make up the structures whereby we safeguard
meaning and communicate it to each other, can only define that
which is known because it is verbally conceived, and which (since
it is inseparable from that act of conception) cannot be known
directly in its essence.

This paradox is inherent in the very nature of the naturalistic
novel, which in order to justify itself as art has to persuade us of
its successful fictiveness (craftsmanship, form, or whatever), and
to make us temporarily accept that it is 'true' on its own terms,
its identity depending on the contrast with what is 'not true'. The
power of novels such as *Madame Bovary* springs from their
hypernaturalness—they are patently fictive verbal constructs, to
be differentiated from the world which recognizes them as such;
at the same time they convince us through conformity with this
very world which guarantees their fictiveness. Their 'truth to life'
induces a mendacious assertion of a shared reality, and they are

only acquitted of such idolatry by the innocent atheism of their refusal to be other than they are. For pure naturalism has in itself no designs upon its readers: it is, essentially, a game.

However, the introduction of a supernaturalist element into the game plays havoc with normal expectation: it is the joker in the pack, its value undetermined. It insists on a corrective to the limitations of naturalism as commonly understood; it declines to repose either in a simplistic materialism or in a materialistic 'spirituality'. Its proffered 'shared reality' extends beyond what is normally accepted as believable: in order to recognize its 'truth' we have to envisage a limitation in the 'not true' that confers on it its identity as fiction; we have, that is to say, to modify our own concepts of what is natural.

For whatever their method of approach, the more serious writers of supernaturalist fiction point by implication towards a spiritual order transcending any human formulations. What Jacques Derrida posits (disparagingly) of negative theology is applicable in their case too:

[It is] always concerned with disengaging a supra-essential reality be-yond the finite categories of essence and existence, that is, of presence, and always hastens to remind us that if we deny the predicate of exist-ence to God, it is in order to recognize him as a superior, inconceivable, and ineffable mode of being.[16]

Although linguistic philosophy may deny the propriety of such an assumption of ultimate transcendence and of a *telos* or source and goal of all meaning, imaginative writers, whether consciously or not, necessarily proceed on the ground that such a belief is justified. This is not simply an act of faith on their part, it is an inescapable concomitant of the desire to celebrate, protest, and in all manner of ways communicate their thoughts and feelings to each other.

In Charles Williams's novel *The Greater Trumps* there is an account of the metaphysical dance of the elements in the created order, as depicted in the figures on the Tarot pack. At the centre of them is the Fool, the unknown quantity who does not move— does not move, that is, for those impure in vision, who mistake its speed for immobility, and accordingly reify it into an idol, solid and unmoving. But to those whose inward eye is pure, the Fool dances with the rest of the figures, seeming 'as if it were always arranging itself in some place which was empty for it'.[17]

In this novel the function of the Fool is the ever-living, ever-adaptable grace of God-in-Christ; but by analogy one can see the same process at work as the element of the supernatural pervades and adapts itself to the fictions which people invent to meet the various social pressures and preoccupations, the fears and affirmative responses, that inform their lives. Although no human concepts or ideological convictions can be equated with the final resting-place or telos, that that telos is in constant motion, taking each place that is empty for it, does allow one to affirm its validity as a fundamental hypothesis.

By virtue of their subject-matter and of its demands upon their creativity, novelists concerned with the supernatural expose the simultaneous necessity and impossibility of defining absolute truth. They unsettle our understanding of the circumambient non-fictional world which asserts, defines, and qualifies the relationship between fictive and veridical elements within the text itself. But that capacity to unsettle is obfuscated by the contemporary inability to provide the subject with a critical terminology appropriate to it. Failure to address the subject other than by interpreting it as displaced cultural, political, or sexual anxiety merely eludes the challenge, and thus the topicality, of supernaturalism. The end result is muddle.

Mystery and the Supernatural

'I like mysteries but I rather dislike muddles.' Mrs Moore's pronouncement in chapter 7 of *A Passage to India* might stand as E. M. Forster's epigraph for a novel which dramatizes the difference between those two concepts. To the very end of the book its central episode remains obscure. Was Adela Quested assaulted by an Indian in the Marabar caves or was she not? Forster himself, when pressed, declared he did not know—which might be construed as mere teasing were it not more likely that he was rejecting the very premises that dictated the enquiry.[18] For the whole point of the caves is that they render such questions nugatory: they symbolize the unknowable, that mysterious dimension which resists categorization in materialistic terms. The attempt to explain the mystery is thus in effect to deny the mystery, and produces muddle—as it does in Forster's novel. Adela withdraws the charge against Dr Aziz, not because she can offer an alternative explanation for her experience in the cave, but because her mind is, for the moment, free of cant.

Behind the drama lies the difference between oriental and occidental notions of reality. For the orient, the material order has its source in spirit: matter is a phenomenal reality only, real reality lies behind the appearances. This concept, so alien to the Western civilization in which the novel had its birth, is, however, common to all religions, persisting in what Aldous Huxley describes as 'the perennial philosophy'.[19] In the West it manifests itself in the Platonic and Neoplatonic philosophical traditions, whence it issues in the work of the major English Romantic poets, only to be personalized, diluted, and divided in the later nineteenth century. It is a view of life that conceives of God as immanent in a creation of which he is the source and goal, and of humanity as a microcosm of a spiritual order whose operative laws relate to the metaphysical world that lies, as it were, beyond and around the world of sense. This is a religion of space, one that views reality as a whole, rather than a religion of time and sequence.

Western Christianity, the Christianity associated with Rome rather than with Byzantium, has always laid stress on time and history, asserting the sacredness of matter implicit in the doctrine of the incarnation of the Son of God. In doing so it rejects the belittling of the physical order which crippled the social culture of the East; at the same time it has increasingly come to disregard the constitutive role of spirit; while its natural offspring, scientific materialism, no longer viewing religion as an expression of the nature of the cosmos, inevitably reduces it to a subjective therapy with which to confront the problems and contradictions of what it is inclined to term 'the spiritual aspect' of life. But for a religious person spirit is no mere 'aspect'; it is the primary reality, and any discussion of the supernatural which does not allow for such belief and take it seriously is self-stultifying from the start.

The contrast between the exclusiveness of the Christian/Islamic monotheistic tradition and the all-inclusiveness of Hinduism is most subtly portrayed in *A Passage to India*. But the novel does not merely dramatize the distinction between Eastern and Western religions and between mystery and muddle: it also portrays the agnostic Western consciousness when confronted by the inscrutability of the cosmos. It is noteworthy that the enlightened Mrs Moore is no more immune to the echoing vacancy of the caves than is the hyper-cerebral Adela. Both women attempt to rationalize their experiences and in so doing bring about a state of muddle. To regard the mystery as though it were a mere puzzle to be solved (which is what Adela's complaint amounts to) is to trivialize it.

'Mystery' is a metaphysical term. Metaphysics, the intellectual exploration of the spiritual, the intangible, and the unseen, is a study necessarily discredited by the presuppositions of linguistic philosophy and scientific materialism; nor is it in much repute in contemporary theology. Nevertheless, once mystery is acknowledged to be more than a definition of human limitation or fallibility, then it becomes a potential object of serious inquiry—as it has been, of course, in every civilization prior to our own, which is unprecedented in having no theological basis for its undertakings, and in consigning the religious and metaphysical understanding of mystery to the category of superstition.

Mystery is not in itself an object of worship; rather it is the dimension in which worship becomes meaningful and all-inclusive. In order to distinguish it from a mere sense of inexplicability, it may be designated 'the mysterium'. This term refers to that condition, or posited spiritual order, to which the sense of religious or mystical awe directs its subject, and into which the participants in the Eleusinian and other mystery cults were admitted; and it is in relation to the mysterium that the concept of the supernatural may most readily be discussed. Whether in the sense of a truth knowable only through divine intervention or of a religious ordinance or rite, the word 'mystery' refers to the eternal source of human life—to what in one vocabulary is signified by 'heaven', in another by the *anima mundi* or place of archetypes. In the teaching of Catholic mystical theology which underlies the Western world's traditional understanding of the supernatural (a theology based not only on personal devotion but also on the spiritual inheritance of Judaism and the cabbala) the mysterium is embodied in the figure of divine Wisdom, at once reflector and articulator of the glory of God. Both theologically and linguistically it is this source of the natural order, and its regulating activity, which should alone be designated supernatural.

The very pattern of Christian worship in the calendar of the Catholic Church proclaims as much: it interprets the earthly events of the life of Christ as 'mysteries', embodiments of what happens eternally in the world of spirit.[20] Of these mysteries, the Transfiguration and the Ascension in particular are concerned with the relation of nature to supernature, representing as they do complementary aspects of the interpenetration of heaven and earth in the person of the divine Word for whom all things exist. They are not preternatural events; they are indices of supernatural reality. The Transfiguration expresses the physical embodiment of that reality, the Ascension the more customary awareness of its withdrawal; but the Ascension is also the reminder of Christ's presence in the heavenly sphere which is the supernatural dimension of this world and which the Transfiguration declares him never to have left. Such a conception is the very basis of the sacramental reading of the material world, and to worry at the problems of a historical ascension is to miss the point of the enacted parable altogether: in this respect twentieth-century materialistic positivism is more spiritually naïve than were the metaphysically attuned responses of previous ages, which possessed an understanding of the supernatural and eternal denied to their more literal-minded descendants.

It is also of the essence of a sacramental perspective that theological truths should be understood as contemporaneous with each other, and time as the sacrament of eternity. Considered theologically, the supernatural is not *part* of nature; it is basic and encompassing reality itself. Accordingly it is a subject for a visionary's attentiveness, not an object of manipulative inquiry, being that order in which spiritual laws of cause and effect operate unqualified and unmediated by physical perceptions. Any use of the adjective 'supernatural' which implies or involves a partial, or a materially dismissive, attitude towards the mysterium really belongs with such terms as 'preternatural' or 'paranormal', terms that are indicative of various responses to the supernatural rather than ontological definitions of it. To substitute one term for either of the others is to confuse the issue: as an authoritative exponent of mystical theology points out,

the forms under which it is natural and necessary for us to conceive of transcendental truths have a real and vital relation to the ideas which they attempt to express; but their inadequacy is manifest if we treat them as facts of the same order as natural phenomena, and try to intercalate them, as is too often done, among the materials with which an abstract science has to deal.[21]

Such an intercalation is what Forster defines, with his character-istic refusal of portentousness, as 'muddle'.

Individuals respond to the mysterium in different ways accord-ing to their individual temperaments and cultural conditioning: in the words of the seventeenth-century Neoplatonist divine, John Smith, 'Such as men themselves are, such will God Himself seem to them to be.'[22] The most primitive and basic of such responses are fear, anxiety, terror, the sense of an approaching menace from outside, a threat to people's security and to what they know and take for granted: it is the experience of the mysterium as *preter*natural, at once dreaded, resented, and (with a seeming perversity) desired.

More affirmatively, the mysterium can be interpreted as paranormal, both as an extension of human powers and as a challenge to them: the reaction to it is one of respectful consid-eration, tempered by a curiosity which finds current practical expression through the deployment of physical energy in the form of technology and science, just as at a more primitive stage such curiosity produced alchemy and, at a lower level, witchcraft and the blatant power-urges of sorcery and voodoo. But for the majority of Western people the mysterium is intuited either in aesthetic contemplation or as a sense of timelessness, or through emotional response to natural forces and the configuration of particular landscapes; alternatively it may be experienced in psy-chological confusion, troubled perception, the world of fantasy and dream.

All such basic responses to the mysterium are expressed ac-cording to each individual's life-experience and attitude to others. They are a vital part of the novelist's material, but always as they interact with the everyday uncontemplative aspect of hu-man life. The point to insist on is that the experience of mystery is not something departmentalized or eccentric: it is an essential part of the awareness of being alive. For example, the sense of a transcendental timelessness is frequently described by Dorothy Richardson, Virginia Woolf, and John Cowper Powys, all of them novelists without allegiance to any particular church or creed: *Pilgrimage*, *The Waves*, and *A Glastonbury Romance* are signifi-cant texts in this connection. It is not an artifically induced con-dition, and for the subject of it it is overwhelmingly real and asks for no explanation. All-encompassing and not to be mastered, it is in the true sense supernatural. At its highest it suggests what Rudolf Otto, designates 'the Holy';[23] but it is an experience open to anyone who for the moment obliterates the thought of past and future. The opening on to the mysterium can be as homely as a garden gate.

Mystery involves both perceiver and perceived. With regard to the former, it is experienced as encounter; it is the condition of that which cannot be cognitively known, the dimension into which rationalistic investigation cannot, and is not prepared to, proceed. For the 'undiscovered country' is indeed undiscoverable. Human consciousness is enclosed in mystery; but mystery, an absolute for purposes of rational discourse, in theological terms is relative to the absolute transcendence of ultimate being that we call God. The impersonal anonymity of that last word is itself a reminder that the supernatural can only be spoken of in metaphors.

Mystery and the Writing of Fiction

As its etymology suggests, the word 'supernatural' denotes that which is above nature; it refers to an order of being superior to that of nature in its physical constituents, to an order which transcends the course of nature. Theologically considered, the word has a teleological reference, that is to say, it relates the natural to its purpose in the mind of God. Richard Hooker uses the word in this sense when he refers to 'Those supernatural passions of joy, peace and delight'.[24] In this originating mode of discourse, 'supernatural' relates its subjects not so much to the unusual or the strange as to the spiritual when understood in the Platonic sense as the origin of being. All three interpretations of the word ('preternatural', 'paranormal', and 'supernatural' when adumbrated as above) correspond to a particular literary methodology.

Some authors treat the supernatural as being in effect preternatural, by which is meant any kind of physical manifestation not attributable to the known laws of cause and effect, anything that differs from what is natural: they portray the supernatural as being not above nature but as contrary to it. Writers whose understanding is confined to this perspective describe the worlds of spirit and matter as being distinct from each other, even opposed to each other. The mysterium is presented as an enigma,

hostile and unnatural. This is understandable enough, for as
H. P. Lovecraft points out,

Because we remember pain and the menace of death more vividly than
pleasure, and because our feelings toward the beneficent aspects of the
unknown have from the first been captured and formalised by conventional religious rituals, it has fallen to the lot of the darker and more
maleficent side of cosmic mystery to figure chiefly in our popular supernatural folklore.[25]

Such stories of unnatural events are designed to unnerve their
readers, to puzzle or disturb them. As a general rule they are written for purposes of (arguably masochistic) entertainment, the
traditional ghost story being a case in point. The majority of such
stories are patently fictitious. The imaginative fruits of superstition, they appeal to credulous, unexamined, and spontaneous
responses; but by their very nature they do not aim to deceive
beyond the time it takes to decipher them. They are justified by
the underlying scepticism of their readers. At its most serious,
the ghost story can take on the character of an allegory or fable,
in which, while a metaphysical reality is posited, it is for provisional literary purposes only. In the context of fable, the preternatural element itself becomes representative of the natural, Henry
James's 'Owen Wingrave' providing a good example of this process.

The second literary perspective treats the supernatural as being *paranormal*, that is as lying outside the range of ordinary
knowledge not as a matter of kind but as a matter of degree—as
when one reads that 'he seemed suddenly animated with supernatural strength'. James describes the attitude in his preface to
The American: 'The real represents to my perception the things
we cannot possibly *not* know, sooner or later, in one way or
another: it being but one of the accidents of our hampered state,
and one of the incidents of their quantity and number, that particular instances have not yet come our way.'[26] This approach to
the mysterium is rational and scientific; its treatment of the hows
and whys of supernatural manifestations asserts no ontological
distinction between material and spiritual categories. The concept of mystery is reduced to the status of a problem calling for
solution. A well-known nineteenth-century example of such an
approach is Bulwer-Lytton's story, 'The Haunted and the Haunters', a tale whose alternative title 'The House and the Brain' indicates its quasi-scientific attitude towards the workings of things

magical and occult. Although it is more concerned with observation than with philosophical inquiry, its author later tried to
combine the two in his alchemical romance, *A Strange Story*. But
this latter book really belongs to a hermetic tradition, which
takes spiritual categories as being inseparable from material ones,
and does so in an inquiring and at times a manipulative spirit.
One might call this a species of imaginative technology arising
from a concern with magic and power. Of the three literary
perspectives this one is the least often developed, its most enduring manifestations being time-warp and time-travel stories.

A third tradition accepts the supernatural for what it is: it
treats material and spiritual experience as aspects or dimensions
of each other, but as subject to the transcendence of the mysterium. It is this approach that is exemplified in *Wuthering Heights*.
It portrays the supernatural as the true province of the imagination; it does not, as does the preternaturalist romance, regard it
as an intrusion upon physical reality, nor, as does the hermetic
one, as the extension of it. It does not exploit the experience of
mystery or seek to explain it. It studies it. This sacramental tradition, so to call it, is contemplative and visionary rather than
speculative in approach, and issues in the form of parable, a tale
which is at once both simile and metaphor. It is a simile in
relation to the spiritual dimension whose processes it demonstrates (as with 'the Kingdom of Heaven is *like* . . .' in the par-
ables of Jesus); but its character is a metaphor for the physical
dimension in which it shares. Unlike the fable, the parable assumes
the presence of a metaphysical reality, and its function is to
demonstrate the workings of the mysterium as it operates through
physical and moral laws.

To sum up: tales of terror, although a product of superstition,
can at their finest take on the character of allegory or symbolic
fable. The hermetic tradition, and the study of the supernatural
when regarded as paranormal, have as their literary form the
treatise, in which the supernatural is examined either as the
controlling force of life or as one phenomenon among others.
The sacramental tradition, on the other hand, is essentially religious in outlook, and expresses itself as parable or meditation.

The tale of terror has been the most prevalent of these literary
forms: it provides the soil from which the others grow, exhibiting
the supernatural as it is experienced at the most basic level of
human response. The tradition flourishes as a kind of literary

underground; horror fiction, indeed, has become a specialized branch of literary activity, its presiding genius Edgar Allan Poe. The second tradition, less in evidence as the tide of hermetic studies has receded, has fertilized the expanding territory of science fiction, especially in its more fantastic manifestations. The sacramental tradition continues to have its Christian and pagan exemplars, the latter being evident in stories of earth-magic, folklore, and the emanations of landscape. But it is an agnostic tradition which is currently in the ascendant. In introverted form it surfaces in novels that deal with psychological disturbance and private fantasy; and in extroverted form in novels which capitalize on the ludic and artificial aspects of fiction-making, projecting a complex relativity as against the simple absolutes of naturalism. The variety of ways in which the supernaturalist writers in each tradition handle their material relates to the nature of fiction itself, as well as to those differing aspects of human experience that popular consent still designates 'the supernatural'. In the supernaturalist novel, where the question of verification is concerned, content and form are reciprocally illuminating.

2

AN ICONOGRAPHY OF FEAR

'I wants to make your flesh creep,' replied the boy.
(Charles Dickens, *The Pickwick Papers*)

Ghost stories are instantaneous levellers. The most sophisticated company can be reduced to rapt attention when the preternatural comes up for discussion, especially if first-hand experiences are relayed, and even when the cited authority is only the ubiquitous 'a friend'. For verification is merely a marginal portion of the game, particularly when the reported experience is malign or threatening. The basic ingredient in all the storytelling is the equation of the supernatural with the uncanny. Ghost stories of this sort readily command imaginative assent, for they express their author's (and their hearers') submerged or unacknowledged insecurities.

The tension between the desire to arouse belief and the need for verification has formed the principal dynamic in supernaturalist fiction for the past 200 years. No less than their predecessors were the late eighteenth-century Gothic novelists concerned with maintaining an external plausibility, and they partly depended for their power to terrify on the assumption that their works were not necessarily fictional, an assumption continued into the following century. (A story such as Poe's 'The Facts in the Case of M. Valdemar', for instance, could masquerade as a medical document, thus heightening its horror still further.) But the early Gothic supernaturalists were concerned to rationalize as well as to authenticate their material. The agreeable palpitations of Jane Austen's Catherine Morland while reading *The Mysteries of Udolpho* would have been assuaged when that reading came to an end, for its author dissolves the mysteries into puzzles and explains them all away. In doing so she was conforming to the prevalent rationalistic temper, in relation to which all subsequent supernaturalist fiction was to be either a protest or a critique. One way or another, supernaturalist writers are inevitably involved in the nature, possibility, and limitations of the quest for objective truth. An examination of their various approaches and techniques will inevitably share in that concern and even, possibly, illuminate it by tracing the literary results of the interplay between the materialist and idealist perspectives outlined in the previous chapter.

Horace Walpole and Ann Radcliffe

One may as well begin with *The Castle of Otranto* (1764), which by general consent is the first deliberate exercise in what came to be known as the Gothic novel. Its influence was acknowledged early, so that the young Fanny Burney, in her preface to *Evelina* (1778), felt it necessary to warn her prospective readers that her own novel would disappoint those who 'in the perusal of these sheets, entertain the gentle expectation of being transported to the fantastic regions of *Romance*, where Fiction is coloured by all the gay tints of luxurious Imagination, where Reason is an outcast, and where the sublimity of the Marvellous rejects all aid from social Probability'.

It was fourteen years since *The Castle of Otranto*, and fashions were beginning to change, partly in response to Walpole's pioneering work, and arguably still more to that of Clara Reeve's relatively conventional *The Old English Baron* (1777): those whom Hazlitt calls 'common-place critics'[1] are always less ready to praise extravagance than they are to commend restraint. Walpole himself had in any case been sensitive to the ridicule his bizarre story might arouse. Dubious as to its reception, he at first tried to pass it off as a romance of sixteenth-century origin, from the library of an ancient Catholic family in the north of England. Being under civil proscription and unable to hold public office of any kind, Catholics were natural repositories for concepts of the sinister, the mysterious, and unknown; while to the world of Twickenham and the metropolis 'the North' was itself suggestive of remoteness and romance. The tale caught on, and as a result Walpole confessed to authorship in a preface to the second edition, attaching to it a statement of literary intent. The book was designed to blend two kinds of romance: the ancient and the modern . . . Desirous of leaving the powers of fancy at liberty to expatiate through the boundless realm of invention, and thence of creating more interesting situations, (the author) wished to conduct the mortal agents in his drama according to the rules of probability; in short, to make them think, speak and act, as it might be supposed mere men and women would do in extraordinary positions.

The extent to which Walpole succeeded in the latter aim is open to question: his characters speak not so much naturalistically as according to a code of literary decorum already well established.

'Young man,' said she, 'though filial duty and womanly modesty condemn the step I am taking, yet holy charity, surmounting all other ties, justifies this act.' (ch. 3)

Matilda's speech as she releases Theodore from the dungeon in the Black Tower anticipates the rhetoric employed by the heroines of Jane Austen's juvenilia, and was to be echoed by innumerable young women in the novels of Walter Scott and his imitators.

Ironically, Walpole saw fit to apologize for what is today the only believable element in the book—the naturalism of the servants' speeches. This division between the marvellous and the matter-of-fact is endemic to the supernaturalist novel. The latter consistently subverts the former in whatever species of the genre. But the presence of the servants is only one feature among many to raise the question as to just how seriously this tale was meant. The abruptness with which the preternatural events take place lends an air of caprice, and their results are less awe-inspiring than surreal, as in the opening catastrophe when the young heir is extinguished by an enormous helmet. Retrospectively there is logic in the various weird happenings, which would, had their portrayal been less perfunctory, have built up to what would then have been a tremendous climax as the gigantic ghostly figure of Sir Alfonso demolishes the castle ramparts; but Walpole eschews the long passages of foreboding that were to be used to such masterly, if misleading, effect by Mrs Radcliffe in *The Mysteries of Udolpho*. His ghostly events, such as a breathing portrait and images which bleed, seem more akin to such popular ecclesiastical miracles as the liquefaction of the blood of St Gennaro in Naples. At other times, as in Manfred's disintegrating powers of speech in face of his three silent guests, and in his interruption of Friar Jerome after he has ordered him to answer

the summons at the gate, Walpole appears to be deliberately undermining the solemnity of his own occasions.

> 'Do you grant me the life of Theodore?' replied the friar.
> 'I do,' said Manfred; 'but inquire who is without.'
> Jerome, falling on the neck of his son, discharged a flood of tears that
> spoke the fullness of his soul.
> 'You promised to go to the gate,' said Manfred. (ch. 3)

Again one is reminded of the young Jane Austen who wrote *Love
and Freindship*.

But although Walpole's purpose of wedding the natural to the
marvellous may be ill-conceived and, except where the castle
servants are concerned, is largely unsuccessful, the consistency
of the preternatural events does amount to something suggestive
of the supernatural in the full meaning of the term. Moreover,
the author's historical sense serves him well to this extent, that
the obvious parallels between Manfred's plans to replace his wife
and those of Henry VIII to divorce Katherine of Aragon lend
force to the otherwise rather trivial momentousness which the
tale attaches to the handing-down of property. However undeveloped the preternaturalist material may be, the motivation of the
mundane elements in the book would be comfortably familiar to
its early readers. *The Castle of Otranto* is more like a Gothic folly,
the diversion of the rationalist imagination, than a stronghold of
supernatural romance.

It is not surprising, therefore, that although Walpole's most
influential successor was Ann Radcliffe (1764–1823) she should
not, strictly speaking, be a supernaturalist writer at all. Her enormously popular romances are as much exercises in romantic
sensibility as they are tales of mystery and dread. Her use of
preternaturalist motifs was designed for aesthetic effect. She
makes a useful distinction between the deployment of what she
calls 'glooms of Superstition' and 'glooms of Apprehension'.[2] The
former can be dissipated through rationalization; but the use of
the other offsets any depletion of emotional effect this might
have had. Indeed, Radcliffe's mastery of suspense springs precisely from a combination of uncertainty as to the phenomena
experienced and a scrupulous rationality as to the record of them.

Her heart became faint with terror. Half-raising herself from the bed,
and gently drawing aside the curtain, she looked towards the door of the
staircase; but the lamp that burnt on the hearth spread so feeble a light

through the apartment, that the remote parts of it were lost in shadow. The noise, however, which she was convinced came from the door, continued. It seemed like that made by the drawing of rusty bolts, and often ceased, and was then renewed more gently, as if the hand that occasioned it was restrained by a fear of discovery. While Emily kept her eye fixed on the spot, she saw the door move, and then slowly open, and perceived something enter the room, but the extreme duskiness prevented her distinguishing what it was. Almost fainting with terror, she had yet sufficient command over herself to check the shriek that was escaping from her lips, and letting the curtain drop from her hand, continued to observe in silence the motions of the mysterious form she saw. (*The Mysteries of Udolpho*, ch. 19)

'It seemed', 'as if', 'something', 'prevented', 'check', 'mysterious'— the language of qualification combines precision with uncertainty: Radcliffe eschews the wilder exaggerations of supernaturalist rhetoric. The dubieties she manipulates so cleverly are based on a refusal to define them further: the materialistic rational self is teased and disturbed by a positive absence of confrontation. Her method is a sequence of evasions and withdrawals, concluding with long-subsequent explanations.

The Mysteries of Udolpho (1794) is indeed a triumph of naturalism over mere sensation, and a vindication of sound principles, right reason, and common sense over nervous apprehension and superstitious gullibility. Emily wins through, and earns the reward of her lover's hand in marriage, her fortitude and sensibility being contrasted with the credulous excitability of her maid, whose speech and behaviour serve in Walpolian fashion as a corrective to the romantic extravagance of other personages in the story. *The Mysteries of Udolpho*, however much Jane Austen may have ridiculed the taste for which it catered, is in its inculcation of rationality and self-control a novel of which she must surely have approved.

The Monk *and* Melmoth the Wanderer

One year after *Udolpho*'s publication there appeared the following suggestions for a do-it-yourself Gothic novel.

Take,
    An old castle, half of it ruinous.
    A long gallery, with a great many doors, some secret ones.
    Three murdered bodies, quite fresh.
    As many skeletons, in chests and presses.

An old woman hanging by the neck, with her throat cut.
Assassins and desperadoes, *quant. suff.*
Noises, whispers, and groans, threescore at least.
Mix them together, in the form of three volumes, to be taken at any of
the watering-places before going to bed.[3]

However, by the time that prescription was concocted the main propulsion of this particular fashion was beginning to run its course, along with the French Revolution whose horrors its own ones so faintly echoed. But the fictive treatment of the supernatural was to be a permanent legacy; and the more gruesome Gothic school, having already produced one definitive example in Matthew Lewis's *The Monk* (1796) was belatedly to inspire a genuine masterpiece in *Melmoth the Wanderer* (1820) by Charles Maturin (1782–1824). These two novels respectively embody extremes of scepticism and emotional commitment; but excessive though their constituents may be, the creative energy their authors bring to them ensures their credibility at the time of reading. None the less, where *The Monk* is concerned, the nature of this credibility is itself in question. By a confident stroke of imaginative plundering the 19-year-old Matthew Lewis (1775–1818), already steeped in the excesses of German literary Romanticism, here seized on most of the accepted ingredients of the tale of terror and pushed them to the point where his bombarded readers had either to suspend their disbelief or to bounce back into scepticism. Lecherous monks, sadistic nuns, rape, murder, incest, black magic, banditry, lynching, blasphemy, and torture—into such a brew the author also stirs a couple of ghosts, a seductive demon, and the Devil himself, with the result that in this mêlée the preternatural elements are deprived of their power to terrify. The one ingredient lacking in *The Monk* is genuine mystery. The pace of the narrative is such that there is no opportunity for contemplation, even for the contemplation of iniquity. But Lewis does handle his supernaturalist themes with great bravura and a disturbingly perverse intelligence that anticipates much late twentieth-century writing of this kind.

Thus the two contrasting appearances of the Devil have a certain rationality in their deployment. When seeking to trap the monk Ambrosio into pursuing his lust for Antonia, he appears as Lucifer, a beautiful youth—though beautiful with a chilliness that paradoxically makes him more seductive still. On the other hand, when he appears in answer to the imprisoned Ambrosio's invocation and demands his soul, he is a foul fiend indeed, who provides a commentary on the nature of the hardened Ambrosio's inmost being. Moreover, in asserting his claim to the hapless monk (whom Lewis makes it difficult for one altogether to detest) he is obedient to moral law: there is nothing gloating or gleeful as he carries off his prize. Even so, the physical horrors of the final chapter are excessive. Ambrosio's victimization lacks the moral logic of the punishments conferred upon the damned by Dante in the *Inferno* of the *Divina Commedia*: the Devil is real enough, but the mercy and grace of God remain invisible. Any sense of the genuinely supernatural is lost.

Imaginative logic is more apparent in the portrayal of the two ghosts. That of the murdered Elvira appears to her daughter in a conventional manner, in order to predict and warn. Beyond its impact on Antonia and the landlady (and the effect here is more ludicrous than frightening) there is nothing especially eerie or alarming in the manifestations. Later on, when Ambrosio is compelled to keep a specious vigil in the haunted room, he is in his turn alarmed by the appearance of the ghost—which, however, turns out, by a comic twist, to be merely Antonia's maid. This mistake reverses a more celebrated and dramatic one earlier in the book, when Lorenzo and Agnes plan their elopement. Agnes intends to escape from the castle disguised as the ghost of a bleeding nun, in whom she and Lorenzo disbelieve. But in the event Lorenzo carries off not Agnes but the nun herself. Lewis handles the nightmarish consequences with commendable (and surprising) restraint. Lorenzo is plagued and sickened by the ghost's nightly visitations until her bones are laid to rest. Here the author plays at ducks and drakes with belief and unbelief alike. At the back of all the weird happenings and charnel-house horrors of *The Monk* lies an essentially Augustan sensibility. Lewis's imagination is more akin to that of Byron, in his letters and in *Don Juan*, than it is to that of Coleridge.

Superficially, Byron might be said to be the progenitor of *Melmoth the Wanderer*—this time the Byron who wrote *Manfred*. With this novel supernaturalist fiction comes of age. In an updated version of the Faust legend, Melmoth, a seventeenth-century Irishman, barters his soul for a century and a half of life, a 'posthumous and preternatural existence', with a power to command the elements which renders him superior to the constrictions of time and space. But this particular Faust also enacts the role of Mephistopheles, tempting his victims in order to shed the burden of his isolation. However, no one will pay the price he exacts, the foregoing of their eternal salvation. When the allotted time is fulfilled he is carried off, disappearing into the sea near his ancestral home.

His intended victims are themselves innocent sufferers at the hands of various forms of human avarice, cruelty, and perverted religious belief; and Melmoth exploits their vulnerability, offering them deliverance at the forbidden price. What makes the situation moving is that the faithful sufferers have little or nothing to justify their making the stand they do, save for their trust in a God behind the appearances. Their oppressors are the idolators, the literal-minded, the 'heartless and unimaginative' who 'are those alone who entitle themselves to the comforts of life, and who alone can enjoy them'.

As for Melmoth himself, his knowledge and power have brought him nothing but disenchantment and a bitter awareness of human frailty. Dialectically the high point of the book comes when he confronts the Spanish maiden Immalee. Immalee has in infancy been marooned on an idyllic Indian island and grown up in solitary innocence, a Miranda without a Prospero. Melmoth tells her of the various human religions, of which only the Christian, in its simplest form, conforms to the beauty she already knows. Melmoth's own knowledge makes him aware with anguish of the blindness to human iniquity implicit in the girl's innocence: he forces her to see the cruelty and horror endemic in life. An enraged and disappointed pessimist, he exists in despair and is, even despite himself, powerless to believe in an alternative to the powerlessness of good; and this spiritual impotence only furthers the work of evil. But he sees through humanity's horror of the Devil.

Enemy of mankind? . . . Alas! how absurdly is that title bestowed on the great angelic chief,—the morning star fallen from its sphere! What enemy has man so deadly as himself? If he would ask on whom he should bestow that title aright, let him smite his bosom, and his heart will answer,— Bestow it here! (ch. 28)

The theme is handled with considerable subtlety, making play with time-shifts and implied narrative with an almost twentieth-century sophistication. The story opens with the death of Melmoth's descendant, an avaricious Irish landowner, whose nephew and heir first hears of the mysterious, ill-reputed ancestor from the lips of a superstitious housekeeper. The tumble-down house and its elderly retainers are evoked with a robust particularity that looks ahead to the best work of Sheridan Le Fanu. In a decaying and fragmentary manuscript the heir reads about the misfortunes of a seventeenth-century Englishman called Stanton, who encounters the Wanderer under a series of ominous circumstances, and is finally presented by him with the great temptation while imprisoned in a lunatic asylum: he is offered deliverance from threatened insanity at the price of an insanity greater still.

Young Melmoth, haunted by a sense of the Wanderer's presence in the house, is then summoned out to a shipwreck, and is saved from drowning by Monçada, a Spaniard who is himself the sole survivor of the storm. Virtually the whole of the remainder of the novel is taken up by the Spaniard's own narrative, and by the other stories that he tells. He himself has been the victim of religious bigotry and family pride. In the hope of disguising his illegitimacy he is compelled by his parents and their confessor to become a monk. The account of his repeatedly foiled attempts to escape from the monastery, his incarceration in the prison of the Inquisition, and ultimate escape by way of a Jewish underground network, is a narrative of extraordinary power, written with something of Samuel Richardson's specificity and concentration.

Maturin uses the machinery of the Gothic novel as a metaphor for a comprehensive picture of human guilt and redemptive heroism: the preternatural elements are vehicles for moral vision. The novel gains its effects less through a piling up of incident than through an accumulation of detail and a sense of interrelated moral issues. Melmoth himself, although frequently demonic in appearance, seems more melancholy than malignant. It is the melancholy which constitutes the malignancy: he embodies the last enemy, Despair. And yet his reasons for despair are so much in evidence that he elicits a certain sympathy. Ultimately the tale is parabolic in effect. Melmoth is not so much a man as an enduring state of mind.

Powerful though it is, not even the author's burning conviction can carry off all the novel's implausibilities. But he displays considerable literary tact, injecting elements of humour from time to time (the old housekeeper and the Irish sybil at the beginning, and the greedy, easygoing Don José, the Spanish chaplain). He also keeps the preternatural within bounds. Melmoth makes his appearance gradually: his consciousness is only occasionally at the centre of the narrative, and he is known more in his effects on other people than in his inner being. The book is thus a dramatization of states of mind, not simply a catalogue of horrors. For all its sardonic indignation and well-nigh-despairing message (Maturin foresees no ameliorating progress in his world) it is full of compassion, especially for the young people, foredoomed as they are by the ambitions and possessiveness of their fathers. Not only Immalee but Monçada and the Jew Adonijah's young son are the victims of bigotry and moral blackmail. Maturin makes some scathing comments in this connection.

Romances have been written and read, whose interest arose from the noble and impossible defiance of the heroine to all powers human and superhuman alike. But neither the writers or readers seem ever to have taken into account the thousand petty external causes that operate on human agency with a force, if not more powerful, far more effective than the grand infernal motive which makes so grand a figure in romance, and so rare and trivial a one in common life. (ch. 22)

This compassion for the young is more generally diffused in Maturin's reflection on the vulnerability of love in a corrupt materialistic world. 'Immalee learned to weep and to fear; and perhaps she saw, in the fearful aspect of the heavens, the development of that mysterious terror, which always trembles at the bottom of the hearts of those who dare to love' (ch. 18). Melmoth has been 'a terror but not an evil' to mankind. His comments to Immalee on the shortcomings of the human race amount to a withering indictment.

They come . . . from a world where the only study of the inhabitants is how to increase their own sufferings, and those of others, to the utmost possible degree; and considering they have only had 4000 years practice at the task, it must be allowed they are tolerable proficients . . . In aid, doubtless, of this desirable object, they have been all originally gifted with imperfect constitutions and evil passions; and, not to be ungrateful, they pass their lives in contriving to augment the infirmities of the one, and aggravate the acerbities of the other . . . In order to render their thinking powers more gross, and their spirits more fiery, they devour animals, and torture from abused vegetables a drink, that, without quenching thirst, has the power of extinguishing reason, inflaming passion, and shortening life—the best result of all—for life under such circumstances owes its only felicity to the shortness of its duration. (ch. 17)

Maturin was no unworthy fellow countryman of Swift. In a foot-note he rather speciously disowns Melmoth's views as being his own; but his eloquence refutes him. None the less, when Immalee replies to Melmoth's contention that 'ten thousand lives a-day are the customary sacrifice to the habit of living in cities' with the words 'But they die in the arms of those they love', he 'was too intent on his description to heed her' (ch. 17).

Despite the extravagance of the central concept, the preter-natural elements play only a supporting part in the presentation of the theme: they are means to an end. Gestures are made to-wards securing a routine imaginative response; and these are at times effective, as in the account of Isidora's journey with Melmoth to the ruined monastery where their marriage is to be solem-nized by a dead monk, or in the repeated references to Melmoth's burning eyes. But the power of his manifestations to his pro-posed victims is mental rather than preternatural: the terror he brings is the state of mind he induces, and its actual worsening of the conditions against which it purports to afford protection. Yet he remains a tragic figure. He, the great deceiver, is himself the undeceived.

Melmoth the Wanderer is almost alone among nineteenth-century novels in indicating a supernatural dimension and signifi-cance transcending space and time. Static and wooden though most of the characterization is, inflatedly rhetorical though much of the language sounds, it remains a powerful expression of ro-mantic protest that verges on theodicy. But the literary tradition to which it belongs was discredited by the time it appeared: Lewis's cerebral disbelief was to prove more normative than were Maturin's visionary intensities. The concepts of good and evil that underlie the supernaturalist tradition called for less outland-ish embodiments: emotion was required to replace sensation as a counterweight to materialist philosophies.

Victorian Gothic: Sheridan Le Fanu

The English Romantic poets, notably Coleridge, Wordsworth, and Shelley, were to articulate discursively what Blake, virtually un-heard, had stated declaratively—that the shaping imagination which interprets the collective representations is human nature's essential characteristic, the basis of wisdom and physical ex-perience. But what was acceptable in poetry was not so readily assimilated in prose fiction, which remained, as perhaps it was its nature to do, committed to a rational materialistic reading of the phenomena of daily life. In terms of the new scientific philosophy human beings were animate recipients of impressions from a world of autonomous inanimate objects; possessed of, and by, emotions, but determined in behaviour by the codes and economic pressures of social institutions. The fact that even so 'romantic' a spirit as Walter Scott (1770–1832) could subscribe to such a reading of human experience shows how limited was the impact made on their contemporaries by the intellectual beliefs of the major poets of the time—a salutary reminder that revolutionary ideas take long to germinate before they are accepted.

But Scott's own influence can hardly be exaggerated. His reputation extended into Europe, and his combination of historical research with a humane understanding of social and religious forces made him an obvious model for subsequent novelists to follow. He was also the first major writer of fiction to take belief in a supernatural world with genuine seriousness—but only the belief, not that world itself. Curious rather than credulous, his imagination was a gathering ground for a wide variety of responses to the supernatural: he inhabited an imaginative borderland, just as he grew up in a geographical one. His particular genius lay in combining regional and historical elements with folklore and supernaturalist traditions, so as to provide a multi-layered rendering of human life, whose surface detail coexists with an awareness of the operations of the spiritual world. His success in achieving this is very uneven, but the attempt to do so is common to most of the Waverley novels, scarcely one of them being without some element of the marvellous, the uncanny, or the paranormal. Where supernatural phenomena are concerned, Scott never commits himself to outright acceptance: his eighteenth-century reasonableness balances his ardent romanticism, and this balance is held in all his finest novels. *Waverley* (1814) contains an instance of his innate caution, in the handling of Fergus McIvor's vision of the spirit of an ancestral enemy, which appears to each leader of his clan in order to announce the approach of death. Fergus describes the encounter in the stilted literary language Scott reserves for his well-born characters; Waverley's reaction to the story comes closer, one suspects, to Scott's own view: 'Edward had little doubt that this phantom was the operation of an exhausted frame and depressed spirits, working on the belief common to all Highlanders in such superstitions' (ch. 54).

Indeed, Scott's attempts to externalize this kind of experience are seldom convincing, and only succeed in short-story form. His most obvious failure is the White Lady of Avenel in *The Monastery* (1820), in whose delineation he allows his reading of German romance literature to overcome his native imaginative tact and sense of local relevance: the comparison with La Motte Fouqué's *Undine* (1811) is highly damaging to him. The White Lady is picturesque rather than portentous; she pipes ditties, and intervenes in the action to farcical rather than to serious purpose, and is finally pensioned off like a family retainer.

A still more instructive failure is Norna of Fitful Head in *The Pirate* (1822). This novel is especially rich in local folklore and depictions of landscape—in this case that of the Shetland Isles. But it carries a tonal ambiguity. Scott wants both to arouse wonder and awe and to provide a preternatural chill, while at the same time endorsing a reductive and sturdy rationalism. The sisters Minna and Brenda Troil illustrate these characteristics, the latter embodying a natural scepticism (and an accompanying timidity in the face of inexplicable phenomena) and the former a romantic passion for Norse mythology which she firmly rejects at the novel's close. But Norna of Fitful Head is more complicated. Her conviction that she possesses magical powers and can control the elements is bolstered by a natural intelligence and ingenuity which causes her to be, in the words of the pirate Cleveland, 'one who knows how to steer upon the current of events'. The near tragic outcome of her belief that she can actually direct them leads to a religious conversion that has been anticipated by the author's own repeated undercutting of her numerous mysterious appearances and interventions. If her irruptions into the narrative are frequently implausible, her personal story is genuinely tragic. Her portrayal is an instance both of Scott's intellectual mastery and of his imaginative limitations. Like the White Lady, she is the product of his reading rather than of his private vision.

His own attitude to Gothic trappings and his understanding of how, from a literary point of view, they can serve the purpose of extending the boundaries of normal consciousness, can be seen in his comments on *The Castle of Otranto*.

He who in early youth has happened to pass a solitary night in one of the few ancient mansions which the fashion of more modern times has left undespoiled of their original furniture, has probably experienced that the gigantic and preposterous figures dimly visible in the defaced tapestry,—the remote clang of the distant doors which divide him from living society,—the deep darkness which involves the high and fretted roof of the apartment,—the dimly seen pictures of ancient knights, renowned for their valour, and perhaps for their crimes,—the varied and indistinct sounds which disturb the silent desolation of a half-deserted mansion,—and to crown all, the feeling that carries us back to ages of feudal power and papal superstition, join together to excite a corresponding sensation of supernatural awe, if not of terror. It is in such situations, when superstition becomes contagious, that we listen with respect, and even with dread, to the legends which are our sport in the garish light of sunshine, and amid the dissipating sights and sounds of everyday life.[4]

The image of the tapestry is ambiguous. Seen from the front, in terms of its finished appearance, it can under certain circumstances induce an illusion of a materialized spirituality; but a tapestry, when seen from behind, is a mass of unrelated and meaningless threads—an analogy of the appearance of the spiritual world when seen without a controlling metaphysic. It is this hinder side of the tapestry which Scott's deployment of superstitious belief and practices resembles.

By the time of Scott's death such concerns were slipping temporarily into the background, and after the accession of Queen Victoria five years later the taste for the Gothic and extravagant receded to the margins of the literary imagination. With one or two exceptions, the major Victorian novelists allowed no more than a peripheral awareness of the supernatural into their work, which was increasingly preoccupied either with the question of social reform, or with recording the comfortable middle-class stability of their own prevailing readership. Ghost stories and tales of terror were provided by way of contrast for such popular magazines as *Household Words*, *The Argosy*, and *Temple Bar* by (to name some particularly skilled practitioners) Charlotte M. Riddell, Amelia B. Edwards, and Mary Elizabeth Braddon, many of whose tales deal with visitations from a past in which romance was now regarded as endemic. Although told with persuasive detail, their stories tended to be rooted in the purely preternatural. The strange events they describe are merely intrusions on ordinary life, and, as such, good for no more than an agreeable shudder. Essentially diversionary, these tales were to reach their apogee as a genre in the ghost stories of Montague Rhodes James.

James's own acknowledged master was the Irish novelist, Joseph

Sheridan Le Fanu (1814–73), the supreme master of the Victorian tale of terror, in his depiction of ominous symbolic landscapes more powerful even than Wilkie Collins or Charles Dickens himself. Dickens's own work in this field not only anticipates Le Fanu's, but provides a notable instance of how supernaturalist material can be incorporated within a narrative concerning the world of everyday affairs. Such an ability is more central to his achievement than is often allowed—and this despite the fact that the most famous of English ghost stories is probably *A Christmas Carol* (1843). Dickens's handling of the supernatural is very much his own; but in this novella 'the Inimitable' did himself imitate, drawing freely on the Gothic tradition as the ghost of Marley enters, dragging his chains behind him—or rather (and here is the Dickensian touch), with those chains wound round him 'like a tail'. And these chains are made up of 'cash-boxes, keys, padlocks, ledgers, deeds, and heavy purses wrought in steel'. The preternatural is hereby rendered logical: Marley in death is as he was, spiritually, in life. And the subsequent detail imposes a scientific precision on the apparition itself, giving a sense of the paranormal. 'His body was transparent; so that Scrooge, observing him, and looking through his waistcoat, could see the two buttons on his coat beyond.' This particularity is as much risible as frightening: as so often with Dickens, hysteria is not far away. The humour, which in *A Christmas Carol* is genuinely frightening (the preceding discovery of Marley's face on Scrooge's knocker being a case in point), can also through its very comicality give birth to what one can only think of as the instant preternaturalization of a basically sceptical tradition. (The comic ghosts of Richard Harris Barham's *The Ingoldsby Legends* (1840) provide some other contemporary examples.) But comedy is present at a deeper level. What Scrooge sees is relative: the Ghosts of Christmas Past and Present show him things as they are; but that of Christmas Yet to Come is disproved by Scrooge's own choice— by his repentance and by the fact that Tiny Tim 'did *not* die'. Thus at one level the preternatural elements are discredited, while the supernatural is asserted through Scrooge's change of heart, which puts him in tune with the workings of Divine Providence. The fiction called *A Christmas Carol* contains, as has been pointed out, 'a double fantasy: one fantasy within another'.[5]

Elsewhere Dickens's supernaturalism can serve a didactic purpose; and in this respect his work continues that of Maturin. The last of the *Christmas Books, The Haunted Man and the Ghost's Bargain* (1847), is a variation upon the *doppelgänger* theme. A chemist is haunted by the resentful memory of his betrayal by a friend: the 'ghost', a figure in his own likeness, offers him forgetfulness at the price of inducing it in all those with whom he comes in contact. There is confusion in Dickens's handling of the theme, since the deadening of the heart in others is presented, not as a matter of psychological contamination, but as a crude exteriorized process. Two kinds of supernaturalist fiction are in conflict: Dickens may be anticipating the methods of Stevenson and Henry James, but he lacks the particular kind of finesse required to make such fictions plausible. He is on surer ground in such straightforward tales of haunting as 'To be Taken with a Grain of Salt' or the better known 'The Signal-Man', a story which lends itself to the author's capacity to render even the most prosaic aspects of the contemporary industrial world with a transfiguring sense of the portentous and grotesque: the longer tale of which it forms a part, 'Mugby Junction', is in its earlier portions a masterly evocation of atmospheric, not to say visionary, dreariness.[6]

Dickens's supreme achievement as a supernaturalist is his interiorization of the preternatural, through suggestive colouring and tone: in this respect he takes the Radcliffian school to its logical conclusion. In his mature art, in the great mystery narratives of *Bleak House* (1851), *Little Dorrit* (1857), and *Great Expectations* (1861), he turns plot into metaphor, and contrives 'real life' strange enough to seem like supernatural drama: Krook's spontaneous combustion in *Bleak House* is an obvious case in point. In these novels Dickens presents coincidence and the complexities of imaginative horror with a forcefulness, an extra-perceptual clarity, that enlarges them into something suggestive of the preternatural. Indeed, the direct handling of the preternatural in *A Christmas Carol* is less disturbing than is, say, Pip's hallucinatory vision of Miss Havisham's hanged body 'with but one shoe to the feet'.[7] *Great Expectations* can also accommodate the attitudes of its own readers, Wemmick's 'castle' at Walworth being the epitome of the cosy, defensive, self-sufficient bourgeois family unit which provides so amenable an audience for the tale of preternatural dread. Not the least remarkable aspect of Dickens's genius was his gift for making the Victorian world seem strange to itself by portraying the dehumanizing ugliness and cruelty of the commercialized mechanistic philosophy that dictated its economic life, with such vividness as to make of London itself a spectre fearful enough to haunt the human beings who had their homes there.

Whereas the plots of Dickens's later novels amount to metaphors, those of Wilkie Collins are raised to a pitch of contrivance which suggests predestination. The mechanics of the narratives serve to subdue the independence of the characters to an overall design: mystery opens on to the mysterium. One of Collins's more subtle presentations of overriding destiny occurs in *The Haunted Hotel* (1879), a short novel which has the additional interest of combining supernaturalist motifs with elements of the detective story. It is a perfect amalgam of its author's various literary strategies, and was singled out by T. S. Eliot on account of its central character, the infamous but resourceful Countess Narona. In most respects she is a stock figure out of Victorian melodrama; but, as Eliot observes, she 'the fatal woman, is herself obsessed by the idea of fatality; her motives are melodramatic; she therefore compels the coincidences to occur, feeling that she is compelled to compel them'.[8] What Eliot fails to reckon with, however, is that the Countess already has a sense of foreboding before the murder of her husband: she tries to avoid marrying him, but is propelled into the union by the brother with whom an incestuous relationship is faintly hinted—we are back in the world of the Gothic novelists. Her death is the result of an obsession with the woman whom she has supplanted in her husband's affections, and thus renders the sense of doom a psychological factor such as might be portrayed in a story by Henry James. The sense of predestination is fulfilled in her collaboration in the mystery's solution: it is as though she herself has both written the scenario and then acted in the play; and that scenario actually is written out by herself, following her breakdown and before her death, serving both as a confession and an explanation. The events of the novel are turned into an artefact within the novel: the drama can thus be read as an acting out of a mystery in the liturgical sense. And Collins, having solved the puzzle for us, ends his novel by what in materialistic terms would be an evasion, but in terms of his own imaginative intelligence is a definition that is necessarily oblique. 'Is there no explanation of the mystery of The Haunted Hotel? Ask yourself if there is any explanation of the mystery of your own life and death.' The fact of the question, so the author would seem to suggest, is its own answer.

In the finest work of Dickens and Collins the complications of plot serve as metaphors for the spirit; and they operate as such through an adroit manipulation of their readers' innate credulity. If the world of appearances, of coincidence and chance, is itself so strange, why should not there be a spiritual universe encompassing and directing it? In the contemporary religious imagination Providence was a potent word; and such a sense of predetermining direction is likewise, if more intermittently, apparent in the fiction of Le Fanu, in whose shorter tales it is challenged and dialectically considered in ways that make him one of the more radical and inquisitive of supernaturalist writers.

In a Glass Darkly (1872) exemplifies the tradition in a variety of forms. It comprises 'Carmilla', a vampire tale; 'The Room in the Dragon Volant', a straightforward crime story; 'The Familiar' and 'Mr Justice Harbottle', two cases of revenge from beyond the grave; and 'Green Tea', in which a melancholic clergyman is driven to suicide by a spectral monkey, a creature from his own subconscious activated by an addiction to strong tea. This overlay of the spiritual with the chemical is characteristic of nineteenth-century interests; and Le Fanu links his tales together by presenting them as case histories in the files of Dr Martin Hesselius, a physician and student of the occult. Yet Hesselius's commentaries are so laconic as to be unenlightening, and the stories have no need of his explanations. In Le Fanu's case, the imaginative force breaks out of the pseudo-scientific framework within which he tries to contain it.

There is a pungency about his writing, a specificity and colloquial vigour (it fairly rampages through his novel *The House by the Churchyard*), which successfully avoids the literary pretentiousness of the late Romantic tradition. Le Fanu has a command of pithy dialect that aligns his work with that of Scott, as in 'Madam Crowl's Ghost', where a housekeeper recalls the spirit of her grotesque employer advancing on a closet which secretes the body of a murdered child.

And what sud I see, by Jen! but the likeness of the ald beldame, bedizened out in her satins and velvets, on her dead body, simperin', wi' her eyes as wide as saucers, and her face like the fiend himself. 'Twas a red light that rose about her in a fuffin' low, as if her dress round her feet was blazin'. She was drivin' on right for me, wi' her ald shrivelled hands crooked as if she was goin' to claw me. I could not stir, but she passed me straight by, wi' a blast o' cald air, and I sid her, at the wall, in the alcove, as my aunt used to call it, which was a recess where the state bed used to stand in ald times, wi' a door open wide, and her hands gropin' in at somethin' was there.[9]

This is altogether more substantial, authentic, and convincing than the stage-managed spectres of the earlier Gothic tradition. Le Fanu has a gift of phrase which enables him to achieve sur-real effects in keeping with his material: 'Surlily he turned away at her words, and strode slowly toward the wood from which he had come; and as he approached it, he seemed to her to grow taller and taller, and stalked into it as high as a tree.'[10] Moreover, the very substantial spectres and demonic agencies are themselves products of their victims' own lives and selves. They materialize, so that others are aware of them, and the nightmare world of the mind engulfs the physical ambience: the hauntings become catastrophic vortices down which the victims spiral from panic to disaster.

Good examples of this process are found in the novella *The Haunted Baronet* (1871)[11] and in the better known 'Mr Justice Harbottle'. Both originated in shorter pieces of writing, and the nature of their development is instructive. The first version of 'Mr Justice Harbottle', called 'An Account of Some Strange Distur-bances in Aungier Street',[12] was published anonymously in the *Dublin University Magazine* in 1853. It describes in cold-blooded detail the ghostly manifestations which threaten the sanity and even, it would seem, the lives, of two students who rent apart-ments in the sometime home of a notorious hanging-judge. The story is told in meticulous detail and with ample internal evid-ence for its authenticity. 'Mr Justice Harbottle' takes matters further. Opening with a preternatural manifestation of an ortho-dox, if characteristically substantial, kind, it proceeds to describe how that manifestation came to take place, and how, in its turn, it was the outcome of still further preternatural happenings. The corrupt and savage judge is haunted by the spectre of a man whom he has had executed for forgery and whose wife he has seduced. The hauntings take different forms, and the earlier instances can be put down to dreams and delusions. But the penultimate chapter is headed 'Somebody has got into the house', and not only do a servant girl and the housekeeper (the dead man's wife) behold spectral figures preparing manacles and a rope, but, more nastily still, the figure of the executed man is discovered sitting in a sedan chair by a little girl who is his own child. This crowning horror is recorded with characteristic deliberation.

To her surprise, the child saw in the shadow a thin man, dressed in black, seated in it; he had sharp dark features; his nose, she fancied, a little awry, and his brown eyes were looking straight before him; his hand was on his thigh, and he stirred no more than the waxen figure she had seen at Southwark fair.

The prosaic detail is reinforced by the comment that

A child is so often lectured for asking questions, and on the propriety of silence, and the superior wisdom of its elders, that it accepts most things at last in good faith; and the little girl acquiesced respectfully in the occupation of the chair by this mahogany-faced person as being all right and proper. (ch. 8)

The demure ironic humour only serves to underline the enormity of what is happening. The eventual discovery of the judge's body suspended from the banisters comes as a logical fulfilment: it is the totality of the tale which induces horror, the materialization of the unconscious guilt and unease of the judge, so that the physical world is, as it were, sucked into the spiritual universe, turned inside out, as the judge himself is judged.

Something similar happens in the development of 'The Fortunes of Sir Robert Ardagh'[13] into *The Haunted Baronet*. The former tale maintains a calm, almost documentary approach to its subject. It opens with a reported tradition—a graphic, substantially realized account of the abduction of Sir Robert by some kind of demonic agency. The story is then retold as it actually happened, first substituting for the demon a singularly unpleasant valet, who disappears, and then providing Sir Robert with a non-violent but mysteriously predicted death, one foreseen intuitively by his wife and sister. In *The Haunted Baronet* many of these elements are combined. The scene is moved from Le Fanu's native Limerick to the fictional town of Golden Friars in northwest England; the demonic valet becomes a wronged nephew who inexplicably survives a drowning, and who in due course vanishes; while the deaths of the two baronets remain virtually identical. But the longer tale incorporates local traditions, family wrongs, and the sense of a demonic agency working by methods of its own towards an end beyond the comprehension of its victim. And the word 'demonic' is exact: the inner desires and guilts erupt into the material world as clothing for spiritual beings ready at hand to use them. In *The Haunted Baronet*, as in all Le Fanu's finest work in this genre, the paranormal becomes the supernatural in its darkest manifestation.

The same process is more sporadically at work in 'Carmilla'. Even by the time it was written (it predates Bram Stoker's more famous *Dracula* by twenty-five years) the vampire tale was a well-established literary phenomenon, among the better known earlier examples being J. W. Polidori's 'The Vampyre' (1819), originally ascribed to Byron, the author's friend. Le Fanu's version is rich in implications—characteristically. The vampire, like her victim, the narrator, is a beautiful girl, first seen in a child-hood nightmare, which prefigures the vampire's attack during a visit to the Styrian castle in which the tale is set. But this vampire, while full of the languor and habits of its kind, is motivated by a devouring sexual hunger: the Sapphic overtones of the story are unmistakable. Its setting is conventional enough; but Le Fanu fills it out with quiet domestic details—the narrator has two friendly but contrasted governesses, and the cleverer of the two displays some understanding of psychic phenomena. Even the oddity of Carmilla's arrival at the castle, deposited by her mys-terious mother, is a brightly lit oddity—an overturned coach which has been set upright is found to contain a hitherto unnoticed and evidently malignant negress. This detail, never developed, leaves a sense of irrational disturbance, typical of the author's refusal to tie up the ends of a story too neatly, or to proffer systematic interpretations. In 'Carmilla' the preternatural remains at that level; yet, even so, the nature of the vampire's yearning for her victim internalizes the story in a way entirely alien to more con-ventional treatments of the theme. The ending holds an ambigu-ity, for the narrator, whose name we are never told, is already dead, though whether to become a vampire in her turn is left unclear.

As elsewhere, Le Fanu propounds no comforting other-worldly solutions. Human beings are presented as the sport of their own urges and desires; but these in turn are manifestations of an encompassing spiritual order for which material objects can supply the alphabet. It is noteworthy that in the most effective tales, although the preternatural elements are weird and often gruesome, they emerge in answer to certain human and social situations characteristic of the world their readers know. Thus 'Schalken the Painter', with its appalling suggestiveness as to a

rape from beyond the grave, attains parabolic status when considered in connection with the bestowal of women in marriage
for purely financial considerations. 'Carmilla', too, might mirror
the intense female friendships arising from the nineteenth-
century segregation of the sexes; and 'Green Tea' would be of
uncomfortable relevance to many a sexually suppressed reclusive
scholar. Greed for money takes on a virulent life in 'Squire Toby's
Will', and God knows what childhood nightmare may have been
reanimated, in turn to reproduce itself, in the hideously precise
'Narrative of the Ghost of a Hand' which Le Fanu so cunningly
insinuates into the ramshackle structure of *The House by the
Churchyard*. The spectres he portrays succeed so well in their business of frightening because their creator intuitively understands
the personal and social vulnerability of their haunted victims.
For Le Fanu's art is essentially subversive, refusing consolation,
palliatives, or certainties: once read, a story like 'The Familiar'
dispels any notion that its author is a mere complacent purveyor
of vicarious shudders.

Nor do his novels, shorn of preternaturalist motifs though they
be, afford any modification of this point of view. In the best of
them—*The House by the Churchyard* (1863), *Wylder's Hand* (1864),
Uncle Silas (1864), *The Rose and the Key* (1871)—physical detail
is so graphically described as to take on at times a well-nigh
preternatural intensity, adding a momentousness that suggests a
supernatural universe hidden behind the clouds of dark unknow-
ing; while their complex, riddling plots seem, like those of Wilkie
Collins, by implication to point to the mysterium. So does the
author's readiness to leave the reader to work out the answer to
some of his unsolved mysteries—for instance, the interventions
of the crazy Uncle Lorne in *Wylder's Hand*. And this effect is
furthered by a gloomy grandeur in descriptions of landscape that
recall the best work of Ann Radcliffe. Le Fanu can achieve an
extraordinary musicality, as in the final paragraph of *Wylder's
Hand*.

Some summers ago, I was, for a few days, in the wondrous city of
Venice. Everyone knows something of the enchantment of the Italian
moon, the expanse of dark and flashing blue, and the phantasmal city,
rising like a beautiful spirit from the waters. Gliding near the lido—
where so many rings of Doges lie lost beneath the waves—I heard the
pleasant sound of female voices upon the water—and then, with a sudden
glory, rose a sad, wild hymn, like the musical wail of the forsaken sea:—

'The spouseless Adriatic mourns her lord.'

The song ceased. The gondola which bore the musicians floated by—a slender hand over the gunwale trailed its fingers in the water. Unseen I saw. Rachel and Dorcas, beautiful in the sad moonlight, passed so near we could have spoken—passed me like spirits—never more, it may be, to cross my sight in life.

'Like spirits'—the sense of transmutation is almost complete. This surely is one of the most melodiously elegaic endings that English fiction has to show, its cadence musical enough to echo that of Ruskin in *The Stones of Venice*.

Dracula *and* The Beetle

The last great nineteenth-century tale of terror is *Dracula* (1897) by Bram Stoker (1847–1912). The author, like Maturin and Le Fanu, was an Irish Protestant, and the sinister undertones of anti-Catholic feeling helped to fuel his imagination, as it had helped to fuel theirs, with an obsession with the physicality of the doctrine of transubstantiation and the horror of enclosed spaces, clerical spies, and of the tyrannies supposedly imposed upon the individual conscience. *Dracula* is also a product of its time, exhibiting the contemporary interest in spiritualism and the occult, together with an optimistic and ameliorative philosophy. Whereas Le Fanu's 'Carmilla' ends with a hint of the continuing power of evil, in *Dracula* not only is the vampire Lucy restored to salvation by a rite of exorcism, but even Dracula himself at the moment of his death has 'a look of peace, such as I never could have imagined might have rested there'. The book thus ends on a positive note, the death of the American, Quincy Morris, adding an element of romantic heroism and the epitaph 'a very gallant gentleman'. Indeed, in the joining together of Lucy's three suitors to protect her friend from the vampire's clutches we have that 'band of brothers' touch found also in the novels of the elder Dumas and of John Buchan.

The comprehensive thoroughness of Stoker's approach is evident in the organization of the narrative. Like Le Fanu in a number of his tales, he introduces the preternatural element at once. Jonathan Harker's imprisonment in Dracula's Transylvanian castle is described with traditional folklore in full play. Wolves howling, coffins filled with the bodies of the un-dead, the pointed teeth and burning red eyes of the vampire women, the hideous bat-like descent of Dracula down the castle wall—all are clearly designed to awaken horror through what can only be described as a softening-up process. Following Harker's escape (by a neat if risky logic of 'where the Count can go, I can too') the scene moves to the Yorkshire coast. The meticulously accurate account of Whitby authenticates the arrival of a storm-tossed vessel carrying the vampire's coffin; while journals, letters, and newspaper reports serve to document what follows in a manner reminiscent of the work of Collins. The variety of personal testimony defictionalizes the material, and thus adds to the reader's disquiet. The climax of this section comes with the death of the first of Dracula's English victims, her transformation into a vampire, and her subsequent dispatch by her former lovers and the learned Dr Van Helsing, who is the occultist authority by now pervasive in novels of this type.

The third part deals with the pursuit of Dracula by the four men, joined by Harker and his wife, who becomes the vampire's next victim. This section and the final one recounting the pursuit of Dracula back to his lair belong more to the genre of the straight thriller, and show a slackening of imaginative invention. Familiar romantic attitudes take over, and the shudder of disquiet is lost.

But for all its disappointing conclusion, *Dracula* deserves its popular reputation—and perhaps something more than that. It articulates a whole tradition. Thus Van Helsing's comments about incredulity might serve as the Tale of Terror's epigraph.

Ah, it is the fault of our science that it wants to explain all; and if it explain not, then it says there is nothing to explain. But yet we see around us every day the growth of new beliefs, which think themselves new; and which are yet but the old, which pretend to be young—like the fine ladies at the opera. (ch. 15)

The humorous touch is characteristic: although where women are concerned *Dracula* flounders in the morass of Victorian sentimentality, it also has a touch of the gusty relish for the macabre that one finds in Beckford's *Vathek* and in Le Fanu. Van Helsing's account of King Laugh enunciates a healthy self-protective scepticism: 'Keep it always with you that laughter who knock at your door and say: "May I come in?" is not the true laughter. No! he is a king, and he come when and how he like. He ask no person; he choose no time of suitability. He say: "I am here." ' The

necessarily persistent use of Van Helsing's broken English is itself a case in point. Even when matters are strung up to agoniz-
ing suspense he goes on getting his words and grammar slightly
wrong, inducing an hysterical impatience which only serves to
further the jaunty breathlessness that is as much a feature of this
book as it is of *The Monk*.

Oh, friend John, it is a strange world, a sad world, a world full of
miseries, and woes, and troubles; and yet when King Laugh come he
make them all dance to the tune he play. Bleeding hearts and dry bones
of the churchyard, and tears that burn as they fall—all dance together to
the music that he make with that smileless mouth of his. And believe me,
friend John, that he is good to come, and kind. Ah, we men and women
are like ropes drawn tight with strain that pull us different ways. Then
tears come; and, like the rain on the ropes, they brace us up, until
perhaps the strain become too great, and we break. But King Laugh he
come like the sunshine, and he ease off the strain again; and we bear to
go on with our labour, what it may be. (ch. 14)

Elsewhere Stoker supplies a grim humour of his own, as when
Van Helsing, breaking into the mausoleum where the vampirized
Lucy lies in her coffin, 'opened the creaky door, and standing
back, politely, but quite unconsciously, motioned me to precede
him'. The narrator is then at pains to point out the irony of this,
and the superfluity of doing so only furthers the absurdity.

Another source of humour in *Dracula* arises from the unnatural confusion between physical and spiritual matters which is the
intellectual root of the image of the un-dead. The efforts to contain the vampire Count also confuse the two—even to the magical
use of wild garlic and (supreme absurdity) the consecrated Host.
These disparate efforts are in derisory contrast with the life-in-
death to which they are opposed. And yet they work, have power.

Although *Dracula* is less gruesome than its reputation might
suggest, it does provide several occasions of genuine physical
horror. Some are conventional and obvious:

The tomb in the daytime, and when wreathed with fresh flowers, had
looked grim and gruesome enough; but now, some days afterwards, when
the flowers hung lank and dead, their whites turning to rust and their
green to browns; when the spider and the beetle had resumed their
accustomed dominance; when time-discoloured stone, and dust-encrusted
mortar, and rusty, dank iron, and tarnished brass, and clouded silver-
plating gave back the feeble glimmer of a candle, the effect was more
miserable and sordid than could have been imagined.

The whole proud panoply of Victorian funeral ceremonies here lies low. But what follows is so obviously obscene that one wonders whether the author realized the effect he was producing— or, rather, whether his publishers did.

Van Helsing went about his work systematically. Holding his candle so that he could read the coffin plates, and so holding it that the sperm dropped in white patches which congealed as they touched the metal, he made assurance of Lucy's coffin. Another search in his bag, and he took out a turnscrew. (ch. 15)

This uncertainty of tone induces an unease which reinforces the tale of terror's literary function of subverting security and routine expectations. When Stoker makes Dracula talk, the result is stagey: it takes one back to the demons of 'Monk' Lewis, though with a crude energy that those more elegant malignancies do not command.

You think to baffle me, you—with your pale faces all in a row, like sheep at the butcher's. You shall be sorry yet, each one of you! You think you have left me without a place to rest; but I have more. My revenge is just begun! I spread it over centuries, and time is on my side. Your girls that you all love are mine already; and through them you and others shall yet be mine—my creatures, to do my bidding and to be my jackals when I want to feed. (ch. 23)

It is unfortunate that the author should see fit to conclude this diatribe by making Dracula say 'Bah!'

Stoker pursues his true *métier* when he incorporates Count Dracula into the late nineteenth-century world. On breaking into the empty house in Piccadilly which has become the vampire's lair, Van Helsing and his companions find 'a jug and basin—the latter containing dirty water which was reddened as if with blood'. The horror is total, for here we intrude upon the criminal living alone with his crime. It suggests in a physical image the self-sufficiency of evil, a concept at once self-contradictory and obscene.

This is true to the Romantic tradition, for the vampire is the other side of the Faustian superman who would live forever. Dracula lives in death because he refuses death, and as such he becomes a significant image of the twentieth-century consciousness. With his portable coffins he symbolizes the denaturalized supernatural being, one whom the materialistic consciousness can only see as a problem to be accounted for and dealt with, a puzzle to be solved: as Van Helsing observes, 'In this enlightened age, when men believe not what they see, the doubting of wise men would be his greatest strength.' Dracula is the symbol of a rejected knowledge that has become evil: 'oh! if such an one was to come from God, and not the Devil, what a force for good might he not be in this old world of ours!' Van Helsing's lament, however, is itself a product of the age he criticizes: he is really only offsetting one aspect of the preternatural with another.

For just as *Dracula* marks the end of a tradition, so the role of the supernatural in subsequent fiction remains in doubt. Was there to be a place for it? And what part would it play in the life of readers? What was to be its imaginative function? The story of Dracula exhibits a number of locations for valid supernatural experience and the response to it. There are the forces of nature— these he can harness to some extent, but ultimately they reject him. There is the psyche—he can invade a madman and over- throw the nature of a virtuous woman; but here he is outflanked. There is the realm of religious witchcraft and magic, where he is confronted directly: but in his defeat on such terms as these he takes them with him, for the supernatural can be truly known only as a religious experience, and genuine religion has no truck with magic. Finally there is the realm of paradox and King Laugh; and there he is not operative at all.

The nature of the dilemma becomes still more apparent in a novel contemporary with *Dracula* and which rivalled it in popu- larity, running through fifteen impressions in as many years. *The Beetle* (1897) by Richard Marsh (*c.*1857–1915) is largely forgotten now, but it makes an instructive contrast with its better known contemporary. The author, an old Etonian, went on to become a prolific inventor of thrillers and fantastic tales, and although *The Beetle*'s subtitle is 'A Mystery', the term 'enigma' fits it better. There is little of the genuinely supernatural in the story. What it does illustrate is the transformation of the august dreams of magic into a threat to the world of materialistic aspirations and achievements out of a forsaken past.

Its setting is that *fin de siècle* London of politicians, heiresses, hansom cabs, and electric telegraphs familiar to readers of the Sherlock Holmes tales of Sir Arthur Conan Doyle; but still more pervasive as a presence is that hinterland of half-developed sub- urbia already hauntingly described by Dickens in *Dombey and Son*

(1848), by Collins in the underrated *Basil* (1852), and by Marsh's contemporary, Arthur Machen. In each case it forms an effective symbol of the underlying emptiness of the materialistic spirit. On a wet night an out-of-work labourer seeks shelter in the downstairs front room of a terrace house, whose window stands alluringly open: once inside, he falls victim to the occupant, an androgynous being who is the priest of a primitive Egyptian cult that sacrifices virgin girls: it manifests itself from time to time as a foul-smelling outsize beetle. The sacred scarab of Isis, it is also 'lamellicorn, one of the copridae', a recognizable species; its onslaughts on those with whom it comes in contact emphasize its repulsive nature.

But the weird 'servant of the Beetle' is itself daunted by a magic it cannot understand: an exhibition of the powers of 'an electrical machine' renders it terrified and subservient. The effect of the incident is to transfer its hypnotic and shape-shifting powers into the category of the paranormal. This book has none of the overriding philosophy or Christian morality of *Dracula*. Instead, it voices an enquiring, materialistic approach to preternatural phenomena that is characteristic of its time. One of the four narrators (Marsh, like Stoker, follows Collins in his careful, veridifying documentation) remarks that

all things are possible I unhesitatingly believe,—I have, even in my short time, seen so many so-called impossibilities proved possible. That we know everything, I doubt;—that . . . our forebears of thousands of years ago, of the extinct civilisations, knew more on some subjects than we do, I think is, at least, probable. All the legends can hardly be false. (ch. 22)

The Beetle, for all its preternaturalistic trappings, remains at the level of the thriller and the horror story. The concluding words make that point decisively: '. . . I am quite prepared to believe that the so-called Beetle, which others saw, but I never, was—or is, for it cannot be certainly shown that the Thing is not still existing—a creature born neither of God nor man.' The phrase is of course a rhetorical flourish; but in view of its theological and philosophical tautology *The Beetle* must, as a supernaturalist story, be regarded as a failure.

As a tale of terror, however, it remains effective, and not simply at the level of inducing a feeling of physical and spiritual unease. The intruder from outside that threatens the comfortable society of the late nineteenth-century world is almost as menacing and omnipresent as it is in *Dracula*. The fear of heathen rites, the suggestion of sexual orgies and of sacrilege against the purity of women, represents a veiled resentment on the part of men against the Victorian moral cultus of woman as the angel mother everlastingly forbidden to man, and everlastingly reproachful of him.[14] The dark gods are viewed as threatening and destructive, but also, very remotely, as vengeful in a collusive way.

As in *Dracula*, so here: three men combine to rescue the heroine from abduction and a hideous death, and here too in the process the author foregoes an experience of supernatural dread for one of immediate suspense. But *The Beetle* is more convincingly rooted in its contemporary world, and there is a lively supporting cast of cabbies, errand boys, landladies, and policemen, who epitomize the rowdy vigour of London life and whose racy dialogue shows up the smart 'literary' talk of their 'betters': the Beetle's neighbour, Miss Louisa Coleman, is as trenchant an old battleaxe as English fiction has to offer. And it is the security of this prosaic encompassing world which guarantees that the malevolent intruder shall have the power to frighten: Marsh gets his finest effects by dint of contrast. No less than in *The Castle of Otranto*, the servants both highlight the nature of the terror and afford protection against its attempts at imaginative hypnotism. Here too King Laugh is, in his way, the master, though of a more ambiguous kind. The preternatural visitant remains at a preternatural level: the fact that old Miss Coleman can observe its comings and goings diminishes that uncanny being to one phenomenon among others. There is no sense any more of supernatural control. The limitations of the Beetle's role in this novel illustrate the dwindling of the Tale of Terror into the temporal restrictions of the ghost story.

M. R. James and the English Ghost Story

By the end of the nineteenth century the tale of supernatural terror was as a mode beginning to fragment, in keeping with the fragmentation of religious belief. In the early writings of H. G. Wells, such as *The Invisible Man* (1897) and many of the short stories, the weird and strange are absorbed into the, by definition, materialistic domain of science fiction. In a parallel vein, in the romances of Henry Rider Haggard, the still only partially explored continent of Africa naturalizes traditional elements of the marvellous in a fusion of realism and fantasy: *King Solomon's*

Mines (1885) and *She* (1887) are celebrated examples. Conan Doyle's Professor Challenger stories, such as *The Lost World* (1912), perform a similar function. Supernatural terror was to find its most distinguished literary expression henceforward in the ghost story, a genre which has many notable practitioners but equally a variety of (the word is irresistible) manifestations. Ghost stories, so called, are not of one species, and the various authors wrote many different kinds of tale, as an anthology like Montague Summers's much reprinted *The Supernatural Omnibus* (1931) makes evident. Several writers in other fields contributed notable examples, among them May Sinclair, Oliver Onions, Vernon Lee, E. Nesbit, Osbert Sitwell, and John Buchan, all writing in a manner, and from a viewpoint, peculiarly their own: the ghost story is a liberating rather than a constricting literary form. Sinclair concentrated on parapsychology, Onions on sensory ambivalence, Nesbit on naked horror, Buchan on landscape and folklore, 'Vernon Lee' (Violet Paget) on extravaganzas, and Sitwell on ornamental, picturesque effect (he also has, in *The Man Who Lost Himself* (1930), a *doppelgänger*). But to the extent that an authentic tradition has developed, it has grown up round the work of Montague Rhodes James (1862–1936), whose first collection, *Ghost Stories of an Antiquary*, appeared in 1904.

Provost of Eton and, previously, of King's College, Cambridge, James was primarily and very thoroughly a scholar, the editor of, among other works, the Apocryphal New Testament. He wrote his stories for the diversion of his friends and pupils, and, as one of his biographers has said, they dramatize

with great skill . . . the unlooked for revelation of an alien order of things, of a wholly malevolent Beyond, linked to our world by a perplexing and dangerous logic: a chance word, an unthinking action, curiosity or simply being in the wrong place at the right time, can all spring the trap.[15]

No one has described the particular Jamesian touch of horror more evocatively than H. P. Lovecraft:

In inventing a new type of ghost, he has departed considerably from the conventional Gothic tradition; for where the older stock ghosts were pale and stately, and apprehended chiefly through the sense of sight, the average James ghost is lean, dwarfish, and hairy—a sluggish, hellish night-abomination midway between beast and man—and usually *touched* before it is *seen*.[16]

The lasting appeal of James's stories lies in their distillation of Gothic terrors which are then quietly caused to impinge on a meticulously drawn familiar landscape.

An evening light shone on the building, making the window-panes glow like so many fires. Away from the Hall in front stretched a flat park studded with oaks and fringed with firs, which stood out against the sky. The clock in the church-tower, buried in trees on the edge of the park, only its golden weather-cock catching the light, was striking six, and the sound came gently beating down the wind. It was altogether a pleasant impression, though tinged with the sort of melancholy appropriate to an evening in early autumn, that was conveyed to the mind of the boy who was standing in the porch waiting for the door to open to him.

One notices that the verb 'open' is, with faint menace, in the active voice. Later, this boy is to see the landscape in a more alarming guise.

Still as the night was, the mysterious population of the distant moonlit woods was not yet lulled to rest. From time to time strange cries as of lost and despairing wanderers sounded from across the mere. They might be the notes of owls or water-birds, yet they did not quite resemble either sound. Were not they coming nearer? Now they sounded from the nearer side of the water, and in a few moments they seemed to be floating about among the shrubberies. Then they ceased; but just as Stephen was thinking of shutting the window and resuming his reading of *Robinson Crusoe*, he caught sight of two figures standing on the gravelled terrace that ran along the garden side of the Hall—the figures of a boy and girl, as it seemed; they stood side by side, looking up at the windows. Something of the form of the girl recalled irresistibly his dream of the figure in the bath. The boy inspired him with more acute fear.

The quiet, natural scene, the initial reasonable accounting for the strange cries, and the mention of what is at once a popular boys' book and a tale of someone in total solitude, lead up inexorably to the two apparitions. (Crusoe's terror at the sight of Friday's footprint would also seem to be in mind.) The terseness of the final statement should not blind us to the word 'acute', since we already know that something or someone has been scratching in the night at Stephen's bedroom door.

'Lost Hearts', from which this passage comes,[17] is a story of which James himself thought poorly, perhaps because of its relatively conventional nature: two very specific ghosts return to destroy an amateur diabolist who has taken their lives in order to prolong his own. Usually James avoids any such extraterrestrial logic. He prefers the inexplicable attack, such as that on the

German lawyer in 'Number 13':[18] 'At that moment the door opened, and an arm came out and clawed at his shoulder. It was clad in ragged, yellowish linen, and the bare skin, where it could be seen, had long grey hairs upon it.' James's ghosts are often not ghosts at all, but mysterious, motiveless agencies, revealing themselves through the medium of tumbled bedclothes, an ancient maze, tangles of hair, lumps of heavy damp sacking, a carved wooden bench-end sprouting fur. One story, 'The Residence at Whitminster',[19] ends on a note of mildly sadistic glee: 'And so . . . Whitminster has a Bluebeard's chamber, and, I am rather inclined to suspect, a Jack-in-the-box, awaiting some future occupant of the residence of the senior prebendary.'

The success of the stories does not, however, lie simply in the ingenuity with which the author succeeds in frightening his readers; rather, it is in the solidity and apparent security of their settings. Some of the early tales take place in Europe, but more usually James sets the scene in landscape of the gentle and exposed East Anglian or Midland sort. The learning cited and displayed lends credibility to the spectral visitations, the dry humour is reassuring, yet often prefaces a ghoulish pounce: James victimizes his characters with tight-lipped aplomb. This background has been well described by Michael Cox as

the enclosed and privileged world in which James—and many of his friends—lived, a world bounded in his case by Cambridge and Eton, country houses and cathedral closes, museums and libraries, and by the holiday points of call—East Anglian inns, comfortable continental hotels, country railway stations—that this bachelor don regularly encountered during the palmy quarter century before the First World War.[20]

This world, reflected so faithfully in the diaries of his friend A. C. Benson, was under a mounting threat. It would seem that James himself had forebodings as to what might happen to it, and reflected them in his peculiarly English way. But what gives his stories their particular sting is the sense that James's imagination is somehow in sympathy with the spectres it evokes. There is a faintly suicidal relish about the way he conjures up these images of decay and menace, a whiff of decadence about his cultivation of unease.[21]

He was to have many imitators, several of them, such as the three Benson brothers and a later writer, L. P. Hartley, bachelors of the bookish kind, readily beset by those fears which are the goblins of the solitary life, and thus all the more susceptible to the creakings of a society ominously on the point of change. Both Hartley and E. F. Benson possessed imaginations fruitful in gruesome imagery; and their work in this kind marks the tailing-off of the Gothic tradition in one of its most popular forms. Nevertheless this version of it persists to the present day: as recently as 1987 there appeared an anthology of ghost stories avowedly inspired by M. R. James.[22] So long as there is a reason-ably secure bourgeoisie domesticated in private property as snugly embattled as Wemmick's Walworth castle, there will continue to be a market for allegorical enactments of its undermining.

It is appropriate that *The Castle of Otranto* should have been the brainchild of the master of Strawberry Hill: both the tale and the house were designed to ornament and amuse an apparently stable social hierarchy. The tale of terror has by its nature always been something of a literary diversion; and a writer like Maturin, who uses the genre prophetically, is bound to attract fewer readers than do those who, like 'Monk' Lewis, aim principally at entertainment. None the less, the mode has proved to be most effective when, as in the case of Le Fanu, it is deployed by a writer whose genius has found in the terror of the preternatural a means of plumbing depths in the psyche which the writing of mere mystery narratives could not provide. In this respect Le Fanu's achievement reverses that of Dickens, whose plots and settings are so much more unnerving than are his formal ghost stories; but in the work of both these writers, and in that of Collins, Stoker, Marsh, and M. R. James, it is the naturalistic fidelity with which they describe the physical contexts of the other-worldly intrusions which makes those intrusions so effec-tive. And the gradual narrowing of focus that is evident in the geographical movement inwards from the exotic territories of Radcliffe and Maturin, through the historically significant or revelatory landscapes of Scott, to the sheltered manor house or residential street of so many later ghost stories, even while it diminishes the apparent occasion for the onslaught of the un-expected, serves to localize it and thus to render it more ines-capable. It is the specificity of place and of the here and now which makes this kind of terror so subversive. Most English mystery writers are only too ready to endorse Sherlock Holmes's declaration, 'I'm a believer in the *genius loci*.'[23]

Yet *genius loci* is the obverse of those vagrant terrors let loose in the Gothic tradition, whose literary apparatus may reflect unconscious guilt over the spoliation of the monasteries and over the subjection of women and the commercial exploitation of their sexuality, an unease which runs through into guilt over the social exploitation of children amid the squalors of industrialism. No wonder Dickens gothicized London and the urban wastelands, and was concerned with missing, deprived, or assumed identities. The individual's place in the social order was being displaced or overridden. But as a literary mode the Tale of Terror was to outlast its relevance. Its degeneration into a crude and arbitary sensationalism is evidence that preternaturalist material no longer commands imaginative assent. The avenging father figures who stalk the corridors of the unconscious have been dragged out into daylight, and the ancestral castle divided into separate apartments. Now it is the wickedness of civilized humankind which commands our fears—not the ghost in the graveyard but the rapist in the alley—the kind of world depicted with such coolly compassionate exactitude in the crime novels of Ruth Rendell. Terror is now invoked by looking in a mirror, and as an image of subversion the preternatural has been rendered superogatory.

3

WATCHERS ON THE THRESHOLD

Man has made mathematics, but God reality.

(W. B. Yeats, *On the Boiler*)

The more materialistic the outlook, the more readily does it fall prey to preternatural dread. Once the concept of the supernatural has been departmentalized and reified, it becomes unmanageable, as much an occasion for black comedy as for fear—a fact on which the sardonic intelligence of D. H. Lawrence fastened in 'The Lovely Lady', a tale of a selfish woman hounded to death by what she takes for spectral voices. Her gullibility, compounded, as in the case of Collins's Countess Narona, by a guilty conscience, is encompassed by her inventor's scorn: a disagreeable satire on literal-mindedness, the story provides a signal instance of what a truly supernaturalist tale is not.

Lawrence both despised and mistrusted the occult. The sense of hiddenness to which it appealed embodied all that he viewed as rotten in contemporary attitudes to religion and to sexuality; and though he was sympathetic to that yearning for spiritual power and wisdom which motivates hermetic studies, none the less he was sceptical as to the results of such inquiries. Yeats's 'half-read wisdom of demonic images' is a phrase that should warn off the literal-minded. The urge to belong to some body of secret gnosis, in order to assert and define one's own identity, provides frequent instances of how the concept of the supernatural can be distorted by human insecurity. Such aspirations are materialistic at heart, for they attempt the systematizing of the immeasurable. The declaration enshrined in the Athanasian Creed that the Incarnation constitutes a taking of Manhood into God, rather than the conversion of Godhead into flesh, defines precisely the difference between a religious understanding of the supernatural and that held by occultists interested in magic and in manipulative spiritual ceremonies.

While both theologians and students of the occult regard material objects as potential vehicles for the operations of the supernatural, the latter study them primarily in the hope of understanding them and participating in their energy: theirs is a scientific spirituality. Theology, the divine science, is contemplative; it focuses upon awareness of the indwelling divine order which constitutes beatitude. While affirming the hermetic belief in a materially operative immanence, it simultaneously recognizes the utter transcendence of the Absolute—Judaism's gift to Christianity. Failure to distinguish between these two responses, the hermetic and the religious, complicates the reactions of subscribers to materialistic philosophies when confronted, say, with Blake's 'matter-of-fact hold of spiritual things'[1] and the systematic study of occult lore undertaken by Yeats. They necessarily view such preoccupations as either the arbitary adoption of convenient symbolism or simply as plain foolishness. Yeats's ambivalent commitments and, still more, the dazzling certainties of Blake, have become (indeed, always were) sources of embarrassment. Both writers were drawing on bodies of seventeenth- and eighteenth-century esoteric writings which rationalistic philosophy on its own premisses had necessarily to disown. Even late twentieth-century readers, with their inbuilt simplistic reflexes of either/or where spirit and body are concerned (Blake's indictment of Bacon, Locke, and Newton mythologizes the dichotomy), are resistant to these matters being taken seriously; and the results, where literary criticism is concerned, is an evasiveness amounting to an ironic comedy, one that veers towards the tragic.

Frankenstein

For ours may indeed be a tragic age, even when we decline to take it seriously: the self-conscious satire of our contemporaries is no defence against the nemesis of history. Confronted with the ecological results of commercialized materialism, one turns back to nineteenth-century writers with a renewed sense of the continuity of their concerns with our own. Although the Victorian age may not have seen itself as tragic, it did take itself seriously: the tragedy there, perhaps, was that, save in a few dissentient voices (Carlyle, Arnold, George Eliot, Hardy), irony was rarely in evidence as a moral weapon, and its implications not popularly accepted. But such seriousness was not a question merely of dedicated, optimistic resolution: it had darker undertones, for the surface prosperity of the age was recognized by its more prophetic thinkers as being the crust above an abyss of physical disintegration and spiritual chaos. That the situation should be analysed in prose as eloquent as that of Ruskin or Carlyle may only have furthered in their readers' minds an exhilarating sense of the drama of their days; and a rhetorical novelist like Dickens could make his protests so enjoyable that his admirers must have been tempted to overlook the consequences to themselves of the social conditions that were those protests' terrible occasion.

But it was the disturbance of religious belief which proved more immediately unsettling, especially to the queasy puritanism that shackled Protestant and Catholic alike. The tension between reasons for disbelief in the master-God and the penalties attendant on apostasy is illustrated in a dream recorded by John Addington Symonds, who as poet, aesthete, and self-tormented homosexual, was a peculiarly vulnerable subject for such a visitation.

I dreamed that we were all seated in our well-lit drawing-room, when the door opened of itself, just enough to admit a little finger. The finger, disconnected from any hand, crept slowly into the room, and moved about through the air, crooking its joints and beckoning. No one saw it but myself.[2]

The associations with the story of Belshazzar's Feast are subtly denigrated: this is like something out of Le Fanu. The sexual implications of the intrusion are sickeningly clear.

Far from being an age of comfortable spiritual materialism, the Victorian period was one of great mental upheavals in which ancient religious trends and traditions surfaced once again. Rational eighteenth-century Deistic Christianity was subverted not only by the other-worldly fervour of Evangelicalism and the revived Catholic spirituality of the Oxford Movement, but also by the growth of interest in mental aberrations, in dreams such as that endured by Symonds, and in spiritualism, mesmerism, theosophy, and other occult philosophies and pseudo-sciences. The growth in communications and the break-up of static, and therefore stable, social units, together with the continued post-Romantic exaltation of the individual and of individual experience, fuelled what amounted to a general personalized recoil from the implicit mechanism and inhumanity of Benthamite social presuppositions, such a recoil as Dickens expressed in novel after novel, most cogently in *Hard Times* (1852): 'Society, considered as it had been in the Age of Reason, as it was in the scientific age, implied naturalism, rationalism, a fixed code of behaviour; this state of affairs by which the bourgeois made his money and sold his soul through a virtuous cynicism was simply unacceptable.'[3]

The various occult traditions, while they looked back to long-established spiritual systems deriving from the Eastern religions and from the Jewish cabbala, also had a contemporary reference. They included a blend of gnosticism, Neoplatonism, alchemy, astrology, and ritual magic. Theosophy, for instance, could be regarded as one method of turning personal spiritual experience into a scientific method, through the kind of eclectic comprehensiveness that made up Madame Blavatsky's widely read compendium, *Isis Unveiled* (1877), a system that provided a hopeful, optimistic counterpoint to the bleaker implications of scientific evolution. Towards the end of the century the aesthetic movement and the cultivation of a self-referential decadence were further protests against the dead hand of materialism. More systematic and scientific in approach was the Society for Psychical Research, founded in 1882: among its members were to be the Cambridge scholar Henry Sidgwick, the critic Andrew Lang, the politician Arthur Balfour, and the doctor and writer of detective stories, Arthur Conan Doyle—a highly respectable cross-section of the ruling social and cultural élite.

But the foundation of much of this interest had been laid earlier in the century with the Romantic Revival and its cult of various kinds of superman, omniscient sage, and master of the elixir of life. Even the arch-naturalist Balzac had in *L'Elixir de longue vie* (1830) and *La Peau de chagrin* (1831) betrayed an interest in the paranormal; while the remarkable success of Mary Shelley's *Frankenstein or The Modern Prometheus* (1818) was evidence of a similar interest among British readers. Indeed, if *Dracula* has become the most durable myth to arise out of the Tale of Terror school, then *Frankenstein* has surely acquired a comparable status in the field of the occult sciences. And yet the book is hard to classify. Is it to be read as a piece of embryonic science fiction? Or as a philosophical romance? Or as a tale of the supernatural, a parable concerning human nature? The plot contains elements which would lend themselves to each of these interpretations.

Mary Shelley was steeped in the humanistic ideas of her poet husband and of her novelist father, William Godwin; she was emulous as well as discipular when she began to write. Her book is as much of historical as of narrative interest. This was a period when vitalist ideas were subsuming ideas of the supernatural into a belief in the paranormal that formed a stage between the magical doctrines expounded in alchemy and cabbalistical tradition, and the advances of contemporary science. Many eighteenth-century thinkers held that

man is endowed with a special sensorium operating independently of the five senses . . . The sixth sense of mesmerists . . . and the sympathetic nervous system of Romantic vitalists established rapport with nature by picking up impulses from a magnetic, electrical, or vital power diffused throughout the universe.[4]

Central to much scientific and philosophic thinking of the time was the attempt 'to reduce the mysteries of life to one basic principle, to identify a single animating agent that at once sustains life and figures as its chief cause'.[5] It is this attempt which forms the subject-matter of more than one piece of magical and occultist fiction.

Mary Shelley's own intention was clear enough, both as enunciated in her Introduction to the reprint of *Frankenstein* in 1830 and in the original Preface which her husband wrote for the first edition. Initially her aim had been to write a story 'which would speak to the mysterious fears of our nature, and awaken thrilling horror'. A conversation with Byron and her husband concerning the possibility of bringing the dead to life through galvanism resulted in a nightmare, the climax arriving when the experimenter awakens to find a monster by his bedside looking down on him with 'yellow, watery, but speculative eyes'. But as Shelley himself remarks in the Preface,

The event on which the interest of the story depends is exempt from the disadvantages of a mere tale of spectres or enchantment. It was recommended by the novelty of the situations which it developes; and however impossible as a physical fact, affords a point of view to the imagination for the delineating of human passions more comprehensive and command-ing than any which the ordinary relations of existing events can yield.

He goes on to cite the *Iliad*, *The Tempest*, *Paradise Lost*, and other works as conforming to this rule, and in so doing sets out the foundations for much subsequent supernaturalist fiction of a serious kind.

In *Frankenstein* the 'scientific' angle is scarcely laboured: the author wisely eschews details beyond her competence, but the hints she drops as to the physical constituents of the eight-foot high man that Victor Frankenstein creates are macabre enough to be worthy even of 'Monk' Lewis. Moreover it is noteworthy that Victor's research and his discovery of the principles of life arise out of his previous immersion in the discredited magical science of alchemy and the writings of the medieval cabbalists. *Frankenstein* has its roots in occult tradition.

The book is in fact less concerned with the marvellous as such than with the philosophical issues for which the marvellous tale affords expression. Frankenstein himself is clear as to the moral of his story: 'Learn from me, if not by my precepts, at least by my example, how dangerous is the acquirement of knowledge, and how much happier that man is who believes his native town to be the world, than he who aspires to become greater than his nature will allow' (ch. 4). In this respect the author's imagination (as the imagination by its very nature is apt to do) goes beyond what her reason has delimited. For the story is not simply about Victor Frankenstein; it is about the being which—or whom?—he has created. How monstrous is the monster?

The answer would seem to be initially that he is a perversion made out of dead men's bones, and worse; and the fact that his maker has the stomach for the job is in part attributed to a rejection of the occult and the preternatural.

In my education my father had taken the greatest precautions that my mind should be impressed with no supernatural horrors. I do not ever remember to have trembled at a tale of superstition, or to have feared the apparition of a spirit. Darkness had no effect upon my fancy; and a churchyard was to me merely the receptacle of bodies deprived of life, which, from being the seat of beauty and strength, had become food for the worm. (ch. 4)

It is this insensitivity which enables Victor to forget that its ugliness will repel any human response to the monster, whose own potentially good nature is thus queered by its appearance. From this perspective *Frankenstein* may be read as a parable of the unjust creation, a theodicy concerning the intrinsically un-lovable. But this aspect of the book is not developed until the close—and even here the role of the monster is ambiguous.

The connection between creature and creator evolves into the kind of reciprocal relationship described by the author's father, William Godwin, in *Caleb Williams* (1794) as existing between Caleb and his persecutor Falkland. Moreover, the brave project of creating a living being degenerates into the accidental raising of a demon: the natural and the occult are set at odds: 'I consid-ered the being whom I had cast among mankind, and endowed with the will and power to effect purposes of horror . . . nearly in the light of my own vampire, my own spirit let loose from the grave, and forced to destroy all that was dear to me' (ch. 7). The heroics of Victor Frankenstein have become the obscenities of Count Dracula; the monster becomes the image of his creator's baser self. At the book's close, with Frankenstein dead, the mon-ster determines to kill himself, since he has no more reason to live. Having been given an objective status through his conver-sation at Victor's deathbed with the book's primary narrator, the Arctic explorer Captain Walton, he achieves—or recovers—a kind of pathos. His prospective end on a burning pyre upon the ice has a fitting and almost redemptive grandeur. As for Frankenstein, he begins to redeem himself when he accepts re-sponsibility for his creature's deeds, and for 'the first hapless victims to my unhallowed arts'. The language is no longer that of scientific inquiry, but that of the occult. Significantly, the occult is concerned not so much with exploring new knowledge as with the recovery of a lost one: whereas science is open to all, cabbalistic wisdom is the province of the few. But in their late nineteenth-century heyday hermetic studies were hailed as being departments of science, and as such could be regarded as valid subjects for naturalistic fiction—an attempt to heal in literary terms the conflict dramatized in *Frankenstein*.

Edward Bulwer-Lytton

The concern with the paranormal which prompted Mary Shelley's germinating dream and the place of the cabbalistical tradition in the novel which resulted, were accordingly to be developed by several later nineteenth- and early twentieth-century writers. And just as Jane Austen satirized the taste for supernaturalist extravaganzas in her day, so did Thackeray in his. The author of *Vanity Fair* had served his literary apprenticeship in journalism and sharpened his wits on the art of parody. 'Punch's Prize Novelists' (1847) is as instructive about contemporary popular writers as Max Beerbohm's *A Christmas Garland* was to be some eighty years later. Nor did Thackeray outgrow his sense of absurdity: the *Roundabout Papers* of 1862 contain in 'The Notch on the Axe' an amusing burlesque of the kind of hermetic novel made popular by Bulwer-Lytton's *A Strange Story*, which had just completed its serialization in Dickens's magazine *All the Year Round*. The mysterious Mr Pinto, with whom the author dines at the Gray's-Inn Coffee House, turns out to be a deathless figure in the tradition of Maturin's Melmoth, the Flying Dutchman, and the Wandering Jew, as well as being, or consorting with, David Rizzio, Cagliostro, Bluebeard, and Mesmer, the contemporary hypnotist. At the end of this farrago the author wakes up to find 'one of those awful—those admirable—sensation novels, which I had been reading, and which are full of delicious wonder . . . If the fashion for sensation novels goes on, I tell you I will write one in fifty volumes.'

For all Thackeray's mockery, Edward Bulwer-Lytton (1803–75) was the most distinguished nineteenth-century English novelist to write about the occult. That he should have done so is not surprising in view of the fact that he had studied mystical philosophy from his youth; and that, despite his not being a Freemason, he was appointed Grand Patron of the Rosicrucian Society when it was founded in 1866. Although his prose style is now irremediably fustian, he was a clever craftsman, who sought to embody some controlling idea or theme in each novel he wrote. He had an ear for the popular pulse, both responding to, and helping to create, the dictates of contemporary literary fashion. But if there is a touch of the charlatan about Lytton, resulting in the failure of any of his novels to preserve much more than a historic interest, one has to respect the intelligence he brought to bear on his handling of the occult. *A Strange Story* even left its mark on *Isis Unveiled*.

Lytton first treated the subject incidentally in a fashionable melodrama, *Godolphin* (1833) and as the primary concern of a shorter tale called *Zicci* (1838). *Zanoni* (1842) is a development of this story, and is a bridge between the romanticism of *Melmoth the Wanderer* and the more scientific perspective adopted in *A Strange Story*. Zanoni is a Rosicrucian scholar, one of two adepts who have contrived to overcome the hazards lying in the way in their design of prolonging their lives indefinitely. The other one, Mejnour, is an austere man of aged appearance, who in Lytton's allegorical scheme embodies the principle of science, the handsome and youthful-looking Zanoni representing art: the schematization reveals a new, detached seriousness in the handling of supernaturalist themes. The story takes place at the time of the French Revolution, culminating with the overthrow of Robespierre, the apostle of Reason—an event which Zanoni has foreseen. But he himself also perishes on the guillotine, victim to his own decision to forswear power in the name of love, embodied here in the Neapolitan singer Viola, of whom Zanoni writes that 'Such hearts live in some more abstract and holier life than their own. But to live forever upon this earth is to live in nothing diviner than ourselves' (book vii, ch. 3). The novel has a sub-plot concerning the artist Clarence Glyndon, who, as a pupil of Mejnour, through a foolhardy impatience raises the demonic Dweller on the Threshold. Its incursion is the one imaginatively convincing moment in the book, for it shows the impossibility of manipulating the mysterium. Lytton was never merely a credulous devotee of the occult.

Zanoni is speculative and quasi-scientific, voicing its author's own idealist philosophy. Lytton takes as his starting-point the hypothesis that there are invisible constituents of matter. He is explicit as to the limitations of materialism—this is a novel about the nature of the supernatural, not a tale of terror. Even the Dweller on the Threshold is accounted for as the embodiment of fear.

Let me explain . . . the dread conditions of its presence. In coarse excitement, in commonplace life, in the wild riot, in the fierce excess, in the torpid lethargy of that animal existence which we share with the brutes, its eyes were invisible, its whisper was unheard. But whenever the soul would aspire, whenever the imagination kindled to the loftier ends, whenever the consciousness of our proper destiny struggled against the unworthy life I pursued, then . . . it cowered by my side in the light of noon, or sat by my bed,—a Darkness visible through the dark. (book v, ch. 4)

But for Lytton the true magic is benevolent, and 'the humblest and meanest products of Nature are those from which the sublimest properties are to be drawn'.

In the present condition of the earth, evil is a more active principle than good . . . It is for these reasons that we are not only solemnly bound to administer our lore only to those who will not misuse and pervert it; but that we place our ordeal in tests that purify the passions and elevate the desires. And Nature in this controls and assists us; for it places awful guardians and insurmountable barriers between the ambition of vice and the heaven of the loftier science. (book iv, ch. 2)

The novel's conclusion looks towards a reconciliation between science and religion. It is a characteristically nineteenth-century aspiration; but its roots are in traditional occult teaching followed by the Rosicrucians. One finds themes in *Zanoni* taken up ninety years later in the novels of a later Rosicrucian, Charles Williams, who himself perceived the merits of the book, even though he realized that it would find few readers in his own time.

Indeed, the only one of Lytton's occultist tales to prove perennially readable is 'The Haunted and the Haunters', which constitutes a link between *Zanoni* and *A Strange Story*. Based on the haunting of a mill at Willington in Northumberland,[6] it is a description of a night spent in a London house whose spectral inhabitants have driven out all previous tenants. The account is eerie from its deliberation; but the peculiarity of the story lies in the second half, which consists of a lengthy explanation of the phenomena, concluding with the discovery in a hidden room of certain powders and instruments which have helped to produce the ghostly manifestations. The wraiths of the dead, the re-enacting of their crimes, have been raised through the machinations of the living.

There are two versions. In the earlier these experiments are traced to the influence of a Cagliostro-like figure who has succeeded in prolonging his own life. The roots of this conception are in eighteenth-century cabbalistical history, and this part of Lytton's tale derives from such novels as Godwin's *St Leon* (1800) and from *Melmoth the Wanderer*; but in the later version the explanatory matter is more purely 'scientific', and anticipates that preoccupation with the rationalizing of the preternatural which characterizes Lytton's most thorough investigation into the occult, *A Strange Story*. The narrator makes his position clear: 'my theory is that the Supernatural is the Impossible, and that what is called supernatural is only a something in the laws of nature of which we have been hitherto ignorant.' This position is modified in the long novel that follows.

A Strange Story addresses itself to a different world of sensibility from *Zanoni*. If the latter looks back to the Gothic tradition, this novel belongs to the world of Wilkie Collins, and anticipates the occultist preoccupations of such early twentieth-century writers as Arthur Machen and Algernon Blackwood. Lytton himself called it a Romance, on account of its 'heritage in the Realm of the Marvellous'; and his narrator Alan Fenwick, a doctor with a materialistic philosophy, is, through the extraordinary events that befall him, compelled to acknowledge that man is not fully human until he has a sense of the supernatural. Despite its apparatus of scientific inquiry and discussion, the plot is rooted in the Gothic tradition—it concerns a man's quest for the elixir of life. Louis Grayle, a debauchee who renews his youth at the price of his soul, crosses the narrator's path in the guise of a handsome, amoral youth called Margrave. Full of a desire to please, but lacking in all moral sense, with his plausibility, charm and cruelty he is the most original character in the book.

A nineteenth-century version of the necromancer-superman Simon Magus, whose figure stalks the pages of occultist literature from earliest times (he is the subject of a rebuke by St Peter for the attempt to purchase spiritual gifts with money (Acts 8: 18–24) and thus epitomizes the fundamental heresy of muddling the procedures of this world with those of the world of the mysterium) Margrave can project his image from a distance, and can direct the wills of the unbalanced, the immature, and the morally undeveloped in order to carry out crimes for his own ends; he also makes use of Lilian, the doctor's betrothed, whose temperament is excessively passive. The account of Margrave's machinations in an old provincial town (called, in the tiresome habit of the day, 'L—', but quite obviously Lincoln) are lively and convincing; but the concluding events in Australia where his plans are defeated by a forest fire of demonic origin are totally implausible, smothered in pretentious imagery and obfuscated by pretentious dialogue. The whole novel, indeed, is overproduced, overexplicated; very rarely does it arouse any genuine thrill of terror or awe. Lilian is droopy even by Victorian standards, and as usual it is the unsympathetic characters who stay in the mind, notably the town's social arbiter, Mrs Poyntz, who, it is amusing to realize, constitutes Lytton's equivalent of the Walpolian servants.[7]

Lytton's humourless portentousness serves him ill in such avowedly dramatic passages as Margrave's initial unmasking in the town museum by a fellow adept. Margrave cries out,

'Something in this room is hostile to me—hostile, overpowering! What can it be?'

'Truth and my presence,' answered a stern, low voice; and Sir Philip Derval, whose slight form the huge bulk of the dead elephant had before obscured from my view, came suddenly out of the shadow into the full rays of the lamps which lit up, as if for Man's revel, that mocking catacomb for the playmates of Nature which he enslaves for his service or slays for his sport. (ch. 32)

At a later stage there are whole chapters of discussion on contemporary theories as to the nature of preternatural phenomena. They have a certain historical interest, whilst investing the book with intellectual pretensions that it cannot sustain. Lytton wants to provide both a work of sensationalist fiction and a scientific treatise, but the respective methodologies cancel each other out. At the end the narrator is converted to belief in the immortality of the soul. The validity or not of the preternatural experiences is left unresolved.

If all the arts of enchantment recorded by Fable were attested by facts which Sages were forced to acknowledge, Sages would sooner or later find some cause for such portents—not supernatural. But what Sage, without cause supernatural, both without and within him, can guess at the wonders he views in the growth of a blade of grass, or the tints on an insect's wing? (ch. 89)

Although this sounds like a resort to rhetoric, earlier in the novel the author has prepared for such a naturalized supernaturalism. In trance the narrator has a vision of a human brain with a threefold aura, representing body, intelligence, and soul, properties with which Margrave, Fenwick, and Lilian are respectively associated—each one being to that extent unbalanced without the others. Lytton is working towards a harmonizing, synchronized resolution characteristic of theosophical teaching, the tradition exemplified in the mystical writings of Jakob Boehme, Emmanuel Swedenborg, and William Blake. A Strange Story may be a novel which unintentionally subverts itself, but it does avoid muddle, even if it offers no especially persuasive account of the mysterium. This limitation is frequent in hermetic novels. As H. P. Lovecraft observes,

occult believers are probably less effective than materialists in delineating the spectral and the fantastic, since to them the phantom world is so commonplace a reality that they tend to refer to it with less awe, remoteness and impressiveness than do those who see in it an absolute and stupendous violation of the natural order.[8]

Fin de Siècle: *Arthur Machen*

Such a violation was to be actively welcomed by the decadents of the *fin de siècle*. Their outlook is defined by the writer Durtal in J. K. Huysman's notorious *Là-Bas* (1891):

we find today, unexplained and surviving under other names, the mysteries which were so long reckoned the product of medieval imagination and superstition . . . a larva, a flying spirit, is not, indeed, more extraordinary than a microbe coming from afar and poisoning one without one's knowledge, and the atmosphere can certainly convey spirits as well as bacilli. (ch. 15)

In such a passage the occult and the paranormal join forces to provide a new perspective on traditional materialism. By the 1890s supernaturalist fiction, in the wake of *fin de siècle* pessimism and the cult of decadence, had taken an altogether darker tinge. The decade saw a number of notable works of horror fiction. *Dracula* comes towards the end of it, but at its opening there is the enormously influential *The Picture of Dorian Gray* (1891), which confirmed the notoriety of Oscar Wilde. And in 1894 there appeared Arthur Machen's still more sensational *The Great God Pan*. Twenty years after Swinburne's *Poems and Ballads*, 'Our Lady of Pain' and 'the roses and raptures of vice' remained current literary coinage. Such morbid extravagance was an understandable reaction against the drab materialism of nineteenth-century political philosophy: as Huysman's Durtal observes, 'It is just at the moment when positivism is at its zenith that mysticism rises again and the follies of the occult begin.'[9]

The Picture of Dorian Gray, however, cannot be regarded as significant in this respect. In essence a fable, it is ambivalent, even confusing, where the supernatural is concerned. For on one reading the central thematic device, whereby Dorian's portrait bears all the marks of that vicious life from which his own beautiful face remains free for twenty years, is an unexplained preternatural occurrence. It happens in response to his own wish, but the event itself is arbitary. The fact that the painter of the portrait has invested the picture with his own feelings for Dorian would seem to be irrelevant. When a belatedly penitent Dorian slashes the picture with a knife it is his own body which is found stabbed through the heart, while the portrait is restored to its pristine state. But any explanation that the picture's ageing is a hallucination on its subject's part is contradicted by the painter's reaction when he sees it. There is no mystery in the unfolding of this particular narrative, only muddle. The book's background is that atmosphere of combined guilt, fear, and defiant speech and behaviour which exemplified the homosexual consciousness in late nineteenth-century Britain. It is this which intensifies the fable from being something playful like Max Beerbohm's *The Happy Hypocrite* (1897) (which is really *Dorian Gray* in reverse) into something portentous and supposedly supernatural. But the sparkle of the verbal wit upstages the preternatural elements irrevocably.

In contrast, the morbid Satanic overtones of *Dorian Gray* are put to a more searching use in *The Lost Stradivarius* (1895), a novel which both illustrates the infiltration of naturalistic fiction by such esoteric concerns and also demonstrates a dichotomy in the English imagination between domestic values and the more permissive way of life associated with 'abroad'. The author, John Meade Falkner (1858–1932) was a successful businessman (he was for many years Chairman of Armstrong-Whitworth, the armaments firm), whose natural instincts were scholarly, and who ended his days in the cathedral close at Durham as Honorary Librarian to the Dean and Chapter. The antiquarian streak in his imagination puts his writings on a par with those of M. R. James; and it seems inevitable that he should have produced three novels saturated in a sense of place and of the past.

The Lost Stradivarius is the first of these, and is the only one to deal with the supernatural. It is unusual among such narratives in lending a sinister colouring to music. When the student baronet, John Maltravers, first plays a Gagliarda for violin and harpsichord by an eighteenth-century Neapolitan composer he inadvertently summons up the spirit of Adrian Temple, a diabolist who is to obsess him to the point of self-destruction. The first manifestation of the spirit is the creaking of a wicker chair, an incident described with a fine mastery of psycho-sensuous observation.

As he shut the pages a creaking of the wicker chair again attracted his attention, and he heard distinctly sounds such as would be made by a person raising himself from a sitting posture. This time, being less surprised, he could more aptly consider the probable causes of such a circumstance, and easily arrived at the conclusion that there must be in the wicker chair osiers responsive to certain notes of the violin, as panes of glass in church windows are observed to vibrate in sympathy with certain tones of the organ. But while this argument approved itself to his reason, his imagination was but half convinced; and he could not but be impressed with the fact that the second creaking of the chair had been coincident with his shutting the music-book; and, unconsciously, pictured to himself some strange visitor waiting until the termination of the music, and then taking his departure. (ch. 1)

The precision of the writing serves to define the underlying tension between the claims of reason and the imagination which is the mainspring of much supernaturalist fiction of this kind, one might almost say, the orthodox version of the whole tradition.

In due course Maltravers is led to discover Temple's own Stradivarius violin, and his furtive concealment of it marks the onset of his degradation. Lured by his passion for the dead man's face, he restores Temple's villa in Naples, discovers the skeleton of his murdered body, and arranges for its burial. Temple's shadow follows him back to Dorset where he dies, terrified but repentant, on Christmas morning. Though more discreetly than *Dorian Gray*, *The Lost Stradivarius* reads like a metaphor for forbidden sexual experience.

The tale is told by Maltravers' sister Sophia, with an appended comment by his friend William Gaskell: once again one finds the importance of authentification in a supernaturalist narrative. The author exercises restraint over preternaturalist effects, and is the more persuasive for doing so. Only one scene belongs to the Gothic tradition, when Sophia watches her brother as he plays the Gagliarda by moonlight to the portrait of Temple in the gallery of a country house, a haunting picture of solitary obsession. Much is achieved through indirection and the comfortable normality of Sophia's mind and outlook. Both she and Gaskell are strict moralists, full of self-respecting rectitude as well as of affection for Maltravers; and this enables Falkner to avoid the excesses of *fin de siècle* rhetorical suggestiveness. Instead of the 'unnameable' vices of *Dorian Gray* (vices once named have a way of dimming into ordinariness) it is made clear that Maltravers has repeated the Satanic rites learnt from Temple's diary, and as a result has seen 'the *Visio malefica*, or presentation of absolute Evil, which was to be the chief torture of the damned, and which like the Beatific Vision, had been made visible in life to certain desperate men . . . the vision once seen took away all hope of final salvation' ('Mr Gaskell's Note'). One is reminded of the cursed knowledge of the wandering Melmoth.

The achievement of the diabolic vision is linked with unspecified pagan practices; but this weakens the force of the author's parable. Once again a reliance on theory obscures the presence of the mysterium, a recurrent weakness in hermetic fiction. For all its imaginative charm, *The Lost Stradivarius* is encapsulated in its time, and its account of the invasion of the sedate English aristocratic rural scene by a power which represented more than that social world's limited imagination could absorb does not altogether transcend its immediate occasion.

The eruption of pagan occult forces in contemporary society is more crudely, but more effectively, portrayed in the flamboyant early tales of Arthur Machen (1863–1947), in whose work a number of manifestations of the supernaturalist tradition relate to each other in revealing ways. He was born in Caerleon on Usk; and the Monmouthshire landscape was to haunt his imagination for the whole of his writing life. In 1881 he moved to London, hoping to earn his living as a writer. In comprising elements of the preternatural, the occult, and the visionary, his fiction amounts to an epitome of nineteenth-century imaginative writing on supernaturalist themes.

The Great God Pan, together with two stories in *The Three Imposters* (1895) (the so-called 'novels' of 'The White Powder' and 'The Black Seal') are Machen's most lurid contributions to the preternaturalist horror story; and all three contain passages that link them with the traditions of occult science. This is conspicuously true of 'The White Powder', the story of a young man who inadvertently consumes a preparation intended for use at a witches' sabbath, and whose body decomposes under its influence: for sheer nastiness the tale more than holds its own with Poe's 'M. Valdemar'. Machen's story has the edge over the latter by virtue of its preternaturalist elements. Here, as in 'The Black Seal' and *The Great God Pan*, as well as in several of the other stories, he makes use of a folk rumour that the Little People, the fairies of popular tradition, were survivors of a prehistoric race still infesting 'the Grey Hills', a thinly disguised rendering of Machen's native region of Gwent. Such a tradition is drawn upon in a Scottish setting and to somewhat different purpose by John Buchan in an atmospheric story, 'No-Man's Land';[10] but Machen adds to it by making his Little People beings of a different order altogether, demonic and yet primitive. Any intercourse between them and humankind destroys the latter. The theme is conveyed with unpleasant immediacy, since Machen's account of the Little

People and the urge to discover their secret knowledge is couched in a style and language that suggests nothing so much as an adolescent fear of, and curiosity concerning, sex: no wonder that contemporary critics decried the stories as decadent. The metaphors are troubling because of the skill with which they are employed.[11]

Even at his most obvious, Machen was a master of atmospheric prose.

The three friends moved away from the door, and began to walk slowly up and down what had been a gravel path, but now lay green and pulpy with damp mosses. It was a fine autumn evening, and a faint sunlight shone on the yellow walls of the old deserted house, and showed the patches of gangrenous decay, the black drift of rain from the broken pipes, the scabrous blots where the bare bricks were exposed, the green weeping of a gaunt laburnum that stood beside the porch, and ragged marks near the ground where the reeking clay was gaining on the worn foundations . . . [They] looked dismally at the rough grasses and the nettles that grew thick over lawn and flower-beds; and at the sad water-pool in the midst of weeds. There, above green and oily scum instead of lilies, stood a rusting Triton on the rocks, sounding a dirge through a shattered horn; and beyond, beyond the sunk fence and the far meadows, the sun slid down and shone red through the bars of the elm trees.[12]

'Gangrenous', 'scabrous', 'reeking'—the epithets of decay are insistent, but controlled by exact observation and an unerring prose rhythm; there is no sense of a pace being forced. And the compressed similes of 'black drift of rain' and 'green weeping' serve faintly to internalize the description, and prevent it from serving merely as an overemphatic backdrop. Ironically, the friends are not sensitive and sympathetic, as this initial description suggests, but a trio who have just committed a singularly vicious crime.

Machen's stories are set in the aftermath of the Jack the Ripper murders, alluded to in *The Great God Pan*, and they capture most vividly the atmosphere of late Victorian London. Indeed the effectiveness of *The Three Imposters*, despite its blatant implausibility, arises from its blending of atmospheres, that of the wooded valleys of Gwent, and of flaming sunsets (to which the author was especially responsive), with the vastness of London and its sprawling northern suburbs. Machen develops the world of *Dombey and Son* and Sherlock Holmes into something apocalyptic.

The juxtaposition of the two worlds is found at its most haunting in *The Hill of Dreams* (1907).[13] The germ of this novel is autobiographical. Lucian Taylor grows up in a vicarage in the valley of the Usk, and goes to London in the hope of becoming a successful writer; much of the author's own depression and discouragement during his early years has gone into the book. It is told throughout from Lucian's point of view and is, indeed, a very early example of the use of a 'stream of consciousness' technique, but it operates on more than one level. Essentially it is an account of the power of the imagination. Factually the plot describes Lucian's adolescence, his solitary exploration of a romantic landscape, his coming to sexual awareness while lying naked on the summit of an ancient earthwork, his initiation by a local farmer's daughter, his romanticizing and internalizing of that experience, and his imaginative re-creation of the old Roman city of Caermaen as the setting for a dream paradise for himself and his beloved. It also describes his literary ambitions, the purloining of passages of his rejected manuscript by an already successful author, and his hatred for his conventional neighbours and their discouragement of his literary ambitions. Having inherited a legacy, Lucian moves to London, and in the northern suburbs, while endeavouring to compose a masterwork, he collapses under the pressures of discouragement and loneliness, and dies from an overdose of drugs in the house of a prostitute. Although this outcome would seem to be totally negative, the second layer of the story challenges such a verdict.

For *The Hill of Dreams* celebrates the elaborations and intensities of Lucian's imagination. Everything is presented more than life-size. The satirical account of the neighbours is sharp and crude, as an adolescent would envisage them. The experiences on the hilltop and with Annie Morgan are of necessity described so obliquely that they seem to mean less than a present-day reader would suppose, or may on the other hand be themselves heightened by Lucian's feverish mind: in the retrospect of his experiences as he sits dying at his writing table (a piece of lyrical prose almost Proustian in its compass) a more specifically sexual interpretation of events is given. But are these versions to be read as the misreadings of a drugged and collapsing mind? The hallucinogenic nature of the London scenes is very marked. At the end, by a switch of perspective that anticipates the work of William Golding, Lucian's dream woman is seen to be a slut and his completed masterpiece illegible. His dead body is lit by a paraffin lamp, whose flaring light 'shone through the dead eyes into the dying brain, and there was a glow within, as if great furnace doors were opened'. That closing phrase echoes words which open the book, descriptive of the impending sunset as Lucian first responds to the mystery of his native landscape during an evening walk. We have come full circle. The ending may be read either as an affirmation of Lucian's values or as a singularly negative touch of irony.

The Hill of Dreams is something of a *tour de force*, written in a fluent prose with Paterian echoes that imposes a unity on what might otherwise be a book divided against itself. For the landscape is suggestive of both good and evil, is both consolatory and threatening. But if it becomes more threatening as Lucian falls prey to his obsession, so also one comes to distrust the hints concerning witchcraft and demonic forces, as being simply the products of delirium. That delirium has in turn been prepared for by the falsification of Caermaen into a self-indulgent fantasy: Lucian, in his hyper-romanticism, his sense of social ostracism, and sheer loneliness, is already on the road to madness. The extraordinary intensity with which the inner London suburbs are described only furthers this realization. And there is genuine irony here, for Lucian has 'longed with all his heart to escape, to set himself free in the wilderness of London, and to be secure amidst the murmur of modern streets'. The novel leaves one with a sense of ambivalence. The author may believe in the supremacy of the imagination; but its visionary nature lays it open to terror and deceit as much as to beauty. Like the seventeenth-century visionary poets Henry Vaughan and Thomas Traherne, Machen came from the Welsh border country, but he lacks their untrammelled awareness of holiness and glory. He was too much a product of his time.

Although *The Hill of Dreams* is a *fin de siècle* novel, the grace of Machen's prose, the eloquent portrayal of landscape, the touching character of Lucian's old clergyman father, all raise it above mere sensationalism; while the satirical invective is more balanced and better placed than it is in a later novel, *The Secret Glory* (1922). Above all, *The Hill of Dreams* is remarkable for its portrayal of burgeoning adolescent sexuality—whether the author was altogether aware of this or not. The account of Lucian's sensations after falling asleep on the earthwork is specific enough:

[he] recollected that for a few seconds after his awakening the sight of his own body had made him shudder and writhe as if it had suffered

some profoundest degradation. He saw before him a vision of two forms; a faun with tingling and pricking flesh lay expectant in the sunlight, and there was also the likeness of a miserable shamed boy, standing with trembling body and shaking, unsteady hands. (ch. 1)

The confused nature of late nineteenth-century attitudes to sexuality could hardly be better expressed. The passage is illuminating when read in connection with E. M. Forster's *Maurice*, written some fourteen years later.

Much of the novel's interest springs from its juxtaposition of good with evil, a connection still more apparent in the finest of all Machen's tales of terror, 'The White People'.[14] The narrative is supposedly taken from the notebook of a young girl who has committed suicide. In language that is childlike but not childish, it tells in one long paragraph of her growing awareness of a world of magical beings and secret rites, transmitted to her in infancy by her nurse and confirmed by her own exploration of the wild landscape round her home. Incorporated into this account are three tales told to the girl by the nurse, all of which recount the loves of mortals for beings of another order, women wailing for their demon lovers and demonic women desirous of mortal men. The handling of the story owes everything to suggestion: the narrator has no sense of evil or wrongdoing, and summons the spirits to her in perfect innocence. It is this innocence which makes the tale so eerie: one is unable to dissociate oneself from the narrator. Her story is the subject of conversation between two friends who discuss and comment upon it, an authenticating device similar to those employed by Algernon Blackwood and by William Hope Hodgson[15] in two collections of supernaturalist tales, *John Silence* (1908) and *Carnacki the Ghost-Finder* (1913), and which here works well.

The opening discussion puts forward a tenet that underlies all Machen's more thoughtful writing in the supernaturalist vein: 'Sorcery and sanctity . . . are the only realities. Each is an ecstasy, a withdrawal from the common life.' For Machen, ecstasy is the product of great literature and should be the aim of all imaginative writing: his essay *Hieroglyphics* (1902) proclaims the belief with passion. And this belief assumes that that other world is 'here', not 'there'. It is such a conviction that underlies the more positive elements in his fiction, found at their most effective in *The Secret Glory* and *The Great Return* (1915), and first articulated in *The Great God Pan*. There the attempt by a pseudo-scientist to lift the veil between the worlds through brain surgery ends in disaster. The woman on whom he operates sees the Great God Pan, has intercourse with him, bears a child, and dies an idiot. The child (called, significantly, Helen) grows up to spread tragedy and horror wherever she lives; and when finally compelled to take her own life, she undergoes a hideous process of shape-changing and disintegrates into what Machen elsewhere calls 'the black swamp whence man first came'. His point in telling the story is orthodox enough. The Kingdom of Heaven is not to be taken by force: try to do so and the powers let loose become demonic. One character in the story is already aware of this, remembering how he had become conscious that

the path from his father's house had led him into an undiscovered country, and he was wondering at the strangeness of it all, when suddenly, in place of the hum and murmur of the summer, an infinite silence seemed to fall on all things, and the wood was hushed, and for a moment of time he stood face to face there with a presence, that was neither man nor beast, neither the living nor the dead, but all things mingled, the form of all things but devoid of all form. And in that moment, the sacrament of body and soul was dissolved . . . (ch. 1)

This is the god Pan, before whom 'the souls of men must wither and die and blacken, as their bodies blacken under the electric current'. Unfortunately Machen chooses to further his effects in an empty rhetoric of horror that baulks the imaginative challenge he was to meet so much more successfully in *The Hill of Dreams*:

I tell you you can have no conception of what I know, no, not in your most fantastic, hideous dreams can you have imaged forth the faintest shadow of what I have heard—and seen. I have seen the incredible, such horrors that even I myself sometimes stop in the middle of the street, and ask whether it is possible for a man to behold such things and live. (ch. 3)

Machen was hampered by nineteenth-century social and literary conventions; but it remains true that to have named the unspeakable atrocities would have obliterated their glamour if not reduced their loathsomeness: one has but to think of the meticulous publicizing of non-fictional twentieth-century atrocities to see the effect of such openness. But even within nineteenth-century limitations Machen forces the pace. He could have learnt from Henry James and the impeccable, infinitely suggestive restraint of 'The Turn of the Screw'.

In his later works Machen adopts the slightly chatty yet ironically archaic manner that derives from the essays of Charles
Lamb. The fiction he wrote in middle age combines both styles.
The Great Return and *The Terror* (1917), while recording preternatural happenings, do so as though coming from the pen of a
curious reporter; but the language is apt to swell into purple
prose at moments of intensity. This is especially noticeable in the
former story, which concerns the coming of the Holy Graal
(Machen's spelling)[16] to a small coastal town in south-west Wales—
Machen calls it Porth, but it is clearly Tenby. Porth also forms
the setting for *The Terror*, which describes a revolt against humans by the animals, thus anticipating Daphne de Maurier's tale,
'The Birds'. With fewer theological pretensions and a climax of
real imaginative power, it is one of the best things Machen wrote.
Porth likewise appears in *The Green Round* (1933). Here one of
the Little People attaches itself to an unwary visitor and becomes
his invisible companion in London and the source of much disagreeable poltergeist activity. The story is ably told, but the vein
had been thoroughly worked out by now.

The Secret Glory, which had been published eleven years earlier,
combines material from both periods of Machen's writing life.
It was begun shortly after the completion of *The Hill of Dreams*,
and is as little like a traditional naturalistic novel than are its
predecessor and *The Green Round*. Ambrose Meyrick is a boy
marooned in a Midlands public school. As a child he has been
shown by his father a Holy Cup, the legendary chalice of St Teilo
which is kept hidden in a remote farmhouse in Gwent; he is
haunted ever after by the knowledge that there is a reality hidden
from the majority of men and women, and that 'we live in a
world of the most wonderful treasures which we see all about us,
but we don't understand, and kick the jewels into the dirt, and
use the chalices for slop-pails'. Machen here spells out the issues
confronting those who in the materialistic twentieth century are
possessed of spiritual vision:

our great loss is that we separate what is one and make it two; and then,
having done so, we make the less real into the more real, as if we
thought the glass made to hold wine more important than the wine it
holds ... The life of bodily things is *hard*, just as the wineglass is hard.
We can touch it and feel it and see it always before us. The wine is drunk
and forgotten; it cannot be held. (ch. 3, s. iv)

The Secret Glory is a discursive meditation on this belief; but it provides little in the way of dramatic embodiment. Ambrose exists with his dreams and memories; he conforms to the rules of the school—a more crudely observed version of E. M. Forster's Sawston in *The Longest Journey* (1907)—and takes refuge in ironic submission, Rabelaisian jests, and a week in London with an Irish servant girl who once took pity on him after he had undergone a beating. The book finally collapses into a farrago of speeches in which Machen denounces modern materialism and spiritual shoddiness; and Ambrose, in a brief epilogue, is dispatched to the East with St Teilo's cup, in order to place it in safe keeping, and on his way back is martyred by 'the Turks or the Kurds—it does not matter which—'. Ultimately the effect is one of cosy gnosticism, a dangerous kind of awareness to which imaginative religious writers frequently appeal.

None the less Machen is an author who, however obscurely, retains his readers. He is a good instance of the artist of intensely personal, unfluctuating imaginative vision, who records a particular experience of life obsessively, but with urbanity and skill. Even in his more uncertain, relatively impersonal early stories one finds a commitment to put forward an other-worldly interpretation of life. The theoretical occult elements give way in the mature work to a visionary mode whereby matter and spirit are presented as aspects of each other. There are no ghost stories in Machen's fiction, and few preternatural happenings; but he was a true hermeticist and believed that all such events emanate from the mind and soul. Accordingly his best work exhibits a refreshing delight in substantial things, not only landscape and weather but also food and drink. It transcends hermeticism, presenting a view of the world that is basically Christian and sacramental.

Algernon Blackwood and David Lindsay

Machen's undogmatic sense of majestic supernatural energy inherent in the physical world contrasts significantly with that of two other writers of a freely speculative nature, Algernon Blackwood (1889–1951) and David Lindsay (1876–1945). The belief that the material universe is to be understood in terms of spirit was held by both of them with a conviction that tended to

suffocate their efforts to proclaim it. They represent one kind of
extreme in the literary handling of supernaturalist material, an
extreme which narrows rather than enhances their appeal, and
which in Lindsay's case had the effect of reducing his status to
that of the object of a cult.

Blackwood, like Machen, Yeats, and Charles Williams was a
student of Rosicrucianism;[17] its influence on him, however, was
secondary to the experiences in the Canadian backwoods which
produced many of his most impressive stories. He has some claim
to be the first author concerned almost exclusively with the uncanny to win serious critical approval. Beginning with a collection called *The Empty House* (1906), his books deal with many
aspects of abnormal experience, ranging from traditional ghost
stories through novels about childhood and the occult, to the
engaging animal fable *Dudley and Gilderoy* (1929), which describes
the adventures and friendship of a parrot and a cat. In later life
Blackwood became a noted broadcaster of eerie tales, his reputation doubtless enhanced by his strange appearance. In reality
an urbane and benevolent man, his deeply lined face and bald
head made him a potentially alarming subject to encounter in a
frontispiece or dust-wrapper.

Like that of Arthur Machen, his work exemplifies diverse aspects of the supernaturalist tradition. The tale that lends its name
to his first book is a straightforward ghost story. A young man
of limited imagination accompanies his elderly aunt when she
decides to spend the night in a haunted house. The story is an
exercise in pseudo-scientific narrative, and is the more unnerving
for it. The ghosts materialize in plausible and thus alarming ways;
while the pacing, command of detail, and building up of tension
are carefully controlled. Blackwood's interest was in psychic and
paranormal phenomena; he remains level-headed, and in the
early tales eschews the rhetoric which he did not deny himself
elsewhere.

His own adventurous youth (he farmed in Canada, then became near-destitute in New York) is graphically depicted in his
autobiography, *Episodes Before Thirty* (1923), and his tales of the
uncanny, many of them based on his own experiences, have more
adventurous settings than the cosy English world of M. R. James.
Blackwood's speciality was to render exposure as terrifying
as earlier writers had enclosure. While he could write excellent
ghost stories of the traditional kind ('The Listener' and 'Keeping

His Promise' are two much-anthologized examples) his most distinctive imaginative vein is to be found in tales such as 'The Wendigo', 'The Willows', and 'The Transfer', in which elemental natural forces are transmuted into demonic powers, the balance between physical occasion and perceptual response being cleverly sustained. At his best Blackwood effects a modulation of sensibility, creates new modes of feeling, albeit in a somewhat laborious prose. His tales in this kind are leisurely (it is necessary for their effect that they should be) and are designed to satisfy the imagination rather than to frighten.

Blackwood's most famous collection is *John Silence*. Silence is a 'psychic doctor' who may in part derive from Le Fanu's Hesselius, but who more certainly, according to R. A. Gilbert, was a portrait of an unidentified member of the Rosicrucian Order of the Golden Dawn.[18] He is a white magican, and is given well-nigh Messianic status as he presides over five 'cases' that exemplify five different aspects of supernaturalist literature. They are concerned with werewolf folklore, the belief in elementals, the witches' sabbath, the relation of animals to the unseen, and vampirism. In each case the setting of the story is all-important —a Swedish island, a lonely English country house, an old French cathedral town, a villa in Putney, a Moravian school in the Black Forest: each one is meticulously described, so that the spectral manifestations emerge from their environment rather than intrude upon it. John Silence himself provides explications of the phenomena, and in all but one case helps to deliver the victims from the threat of demonic forces. The book is a classic instance of the occultist tradition.

The long short story suited Blackwood's gifts. He is less successful with the novel. *The Centaur* (1911) exhibits both his ambition and his limitations. The story of an Irish wanderer, who comes to realize that his senses and inmost feelings are themselves the expression of the soul of the planet, is related at second hand by a method of indirection clearly influenced by Conrad. Blackwood labours greatly to convey with plausibility the emergence of this man's awareness of a vaster spiritual world, tracing its growth until his mysterious disappearance in the Caucasian mountains. The elemental forces portrayed in 'The Willows' and 'The Wendigo' as being demonic are here benign and of a higher order than mankind: 'the Call of the Wild raised to its highest power . . . the call to childhood, the true, pure, vital childhood of the Earth—the Golden Age—before men tasted of the Tree and knew themselves separate . . .' (ch. 16).

The novel is highly programmatic and burdened with lengthy passages of exposition and analysis; the rationalistic Dr Stahl, who endeavours to interpret O'Malley's experiences as paranormal, lends an air of dialectic tension to the narrative. The German mystical poet Novalis, Henri Bergson, the philosopher of instinct and intuition, and the spiritualist Oliver Lodge are all cited in confirmation of the author's case; but the pervasive influence is that of William James's *A Pluralistic Universe* (1909). The cult of the simple life associated with Henry David Thoreau and Edward Carpenter is raised to a metaphysic; it becomes an agent of Pan-worship, the belief that humanity forms part of the corporate consciousness of the earth. But its heavily assertive, exclamatory language weighs this novel down. Ellipses and capitalizations proliferate: Blackwood's prose is here as fustian as is that of Bulwer-Lytton.

If *The Centaur* founders on didacticism and explanatory matter, *The Human Chord* (1910) is lively enough to suggest a forerunner to the kind of novel perfected by Charles Williams. Blackwood draws on his knowledge of the cabbala and his studies as a member of the Order of the Golden Dawn. Philip Skale, a former clergyman living in a large house among the mountains of central Wales, is experimenting with the occult properties of sound, aiming to discover the precise notes with which to pronounce the Tetragrammaton or Name of God, a preoccupation not unlike that of the contemporary Russian composer Alexander Scriabin with the mystic chord. He gathers three singers around him, with the promise that they shall, in uttering the Word, become as gods. The experiment is a classic blasphemy; and it fails because the love of two of the participants for each other makes them prefer the bodily reality they know to the spiritual immensities that Skale holds out to them. This in turns sets in motion a further cause of failure, the adept's imperfect sounding of the Name; he brings destruction upon himself. The lovers escape, secure in their knowledge that the real power and glory are theirs already.

The book voices, as do those of Williams later, a rejection of occult magic in favour of the sacramental magic of ordinary human life; to subsequent readers there is a certain predictability about it. What makes it memorable is the element of grotes-querie which is a feature of Blackwood's imagination, and which links him to such predecessors as Lewis and Maturin. The Reverend Philip Skale is a robust creation, with his joviality, his thundering voice, and enormous size; there is little underhand or menacing about him, and Blackwood's accounts of his dealings with the diminutive tenor Robert Spinrobin have a good deal of possibly conscious humour.

It is in his tales of forests and of impersonal spiritual forces that Blackwood expresses himself uniquely. For him 'everywhere in Nature there was psychic energy' and the effect of frustrating such energy is powerfully suggested in one of the best of his longer tales, 'The Damned'. But he remains a disappointing writer. His ambition is so great, his intelligence so sure, that he ought to be the master in his particular field. But it is precisely as a writer that he fails, not invariably but often enough to render him a less immediate and significant figure than one might expect. His literary shortcomings are summed up by H. P. Lovecraft (whose own glaring lapses were partially retrieved by the obsessive power of his more limited preoccupations).

Mr Blackwood's lesser work is marred by several defects such as ethical didacticism, occasional insipid whimsicality, the flatness of benignant supernaturalism, and a too free use of the trade jargon of modern 'occultism'. A fault of his more serious efforts is that diffuseness and long-windedness which results from an excessively elaborate attempt, under the handicap of a somewhat bald and journalistic style devoid of intrinsic magic, colour and vitality, to visualize precise sensations and nuances of uncanny suggestions.[19]

It is indeed the limitation of the hermetic tradition that its own preoccupations and perspectives by their very nature, lay it open to such pointed strictures.

A similar criticism might be made of the novels of David Lindsay. Like Arthur Machen, he is a minor writer with a small but devoted following: his work sold poorly and he received little or no recognition for it; the last two of his seven novels were not published until he had been dead for over thirty years. But although it is the product of an obstinate self-isolated vision, Lindsay's work can command respect for its intelligence, even if the author's limited powers of communication have now dated most of it beyond recovery: in Victor Gollancz's words, 'he thought and felt and imagined superbly, but wrote abominably'.[20] However, in both his philosophy and his methodology he is an instructive writer where supernaturalist themes are concerned.

His first and only well-known book, a work of space-travel fiction called *A Voyage to Arcturus* (1920), is a parable of the nature of the religious sense and of the workings of self-deception and illusion. Lindsay's vision is bleak and pessimistic, virtually dualistic. His world is in the power of the arch-deceiver Crystalman; but Muspel, the state of pure joy and truth, transcends totally every embodiment of it, for the knowledge of Muspel is transmitted through the distorting fumes of the self-delighting shape-shifter Crystalman. Lindsay's own reading of life stresses the absolute otherness of the supernatural; accordingly he makes no concession to romance. The bleakness of his outlook is matched by the charmlessness of his style and the sharp crudity of his descriptions. The planetary creatures he has imagined anticipate the still cruder embodiments of ideas and emotions to be found in the space-fantasies of John Cowper Powys, such as *Up and Out* (1957).

By implication Lindsay's point of view places all supernaturalist fiction in the parabolic category: that this should necessarily be so in a world governed by Crystalman is the whole point of *A Voyage to Arcturus*, a work which stubbornly beats off its would-be explicators. In his subsequent novels Lindsay, perhaps chastened by the poor sales of his masterwork, essayed a degree of compromise with traditional expectations of what a supernaturalist novel should contain.

The first of them, *The Haunted Woman* (1922), recounts the meetings between a restless, lonely girl and an older man, meetings which take place in a mysterious suite of rooms in the latter's ancient country house in Sussex. They happen out of time, and in the course of them the couple recognize their spiritual affinity, only to be baffled by the fading of the vision when they are outside the enchanted moment. Lindsay's description of the girl's discovery of the rooms by way of a staircase that is only sometimes visible has a magic comparable with the best work of George MacDonald and Walter de la Mare: there is genuine power to convey what stepping into other dimensions of time and space may mean.

There was revealed a doorway, but no door; another flight of wooden stairs started to go down immediately beyond. Isbel persuaded herself that she would still have time to explore a little.

Half-way down, the hall came in sight . . . She could not understand. Near the bottom she realised that she was coming out by the side of the fireplace—in other words, that this staircase was identical with that by which she had ascended . . . How this could possibly be, however, she had no more opportunity of asking herself, for at that moment she reached the hall, and at the very instant that her foot touched the floor every detail of her little adventure flashed out of her mind, like the extinguishing of a candle. (ch. 5)

The ellipses are the author's. The plainness of the language and the delicate rhythmic phrasing seem to enact the metaphysical displacement.

This verbal exactitude is only maintained in the visionary parts of the novel. These resemble the dream encounters of the lovers in George du Maurier's *Peter Ibbetson* (1892); similarly, the man in medieval garments whom Lindsay's lovers espy in October through a window, playing an ancient fiddle among flowering hawthorn, irresistibly recalls a similar moment in *Ash Wednesday* where

> . . . beyond the hawthorn blossom and a pasture scene
> The broadbacked figure drest in blue and green
> Enchanted the maytime with an antique flute.

Can Eliot have read *The Haunted Woman*?

Lindsay's attempt to incorporate his visionary passages in a tale of contemporary life is not, however, a success. The meticulous social details are inevitably out of date (a perpetual hazard for the ultra-naturalistic novelist), the dialogue painfully stilted. Lindsay was more a visionary than a born writer of fiction: in the words of one of his champions, 'In [his] work life is continually shown as shadow level after shadow level, with each shadow hiding reality from those who move in the world below.'[21] For this very reason the supernaturalist scenes in *The Haunted Woman* are peculiarly effective, and the rather shady (in both senses of the word) mediumistic Mrs Richborough is the most memorable human being the author was to create.

Both Lindsay and Blackwood exemplify the tendency of the hermetic writer to subordinate experience to theory. This weakness is especially marked in Lindsay's later work:[22] *Devil's Tor* (1932), for instance, an account of an attempted revival of the worship of Isis, is obese with long-winded disquisition. And though many of Blackwood's tales of the outdoors have the ring of authenticity and are patently sincere in their attempt (so laborious at times) to describe the indescribable, one is continually aware of a conscientious intellect masterminding the literary project. This robs even his books about children of the appropriate playfulness. And Lindsay is not playful at all: the humorous scepticism which the greatest supernaturalists can on occasion command, is something that he entirely lacks. In supernaturalistic as much as in other imaginative writing this can be a fatal limitation. To fail to see it as such is to stop short of recognizing the supernatural as a category that transcends all ordinary experience and which can therefore never be compared with it in terms of measurement. A decent agnosticism is called for when confronting a mysterium that baffles all attempts at penetration or control. Human imperceptiveness ensures that the division between the seen and the unseen is not a matter for tragedy alone.

Travelling in Space-Time

Certainly a degree of comedy attends the literary fortunes of *An Adventure*. When in 1911 Anne Moberly and Eleanor Jourdain published a pseudonymous account of their extra-temporal experience while walking near Le Petit Trianon at Versailles, in which they identified a particular figure with Queen Marie Antoinette, they caused a stir that was more than simply fashionable: the book went into three editions before being reissued in 1931, with an introduction by the novelist Edith Olivier and a note by J. W. Dunne. Since Miss Moberly was Principal of St Hugh's College at Oxford, and her companion was to be her successor, the two women became controversial figures over the years for those in the know. Such claims to psychic intimations were deemed unsuitable, not to say ridiculous, in academics, and the veracity of the account was subsequently discredited. The episode has become a classic instance of its kind, for whatever the conclusive explanation of the story (and in this sort of case there can be none) the curiosity it aroused highlights the paradoxical response to such occurrences—an appetite for preternatural phenomena and a simultaneous one for having those phenomena explained away.[23]

Concern with the occult took on a more scientific aspect as the twentieth century proceeded, a good deal of interest being shown in various notions and philosophies of time. J. B. Priestley's play *Dangerous Corner* (1932) and its two successors enjoyed a considerable popular success, as did Dunne's own treatise *An Experiment with Time* (1927), with its theories as to temporal relativity, and the synchronistic nature of psychic experience—a rewriting in scientific language of traditional esoteric doctrines concerning eternity and the world of spirit. Earlier in the century time-travel had been a matter of either going backwards (as in Ford Madox Hueffer's *Ladies Whose Bright Eyes* (1911), in which a businessman finds himself transplanted into the Middle Ages) or forwards, as in some of the scientific romances of H. G. Wells: both kinds tended to serve as vehicles for social commentary and satire.

An interesting case of the use of the paranormal as an ingredient in fictive naturalism occurs in William Gerhardie's *Resurrection* (1934). This book is a teasing semi-autobiographical account of an experience of astral-projection. It concerns itself in part with the sceptical or uninterested response of the narrator's hearers; significantly, the apathy is directed less at the phenomenon itself than at his interpretation of it as valid proof that death is an illusion. A perfect instance of a supernaturalist novel written in a materialistic vein, *Resurrection* contains enough debatable matter, and enough matter made up of debate, to induce scepticism of scepticism itself and of the author's (and by implication his readers') own readiness to give credence to so apparently simplistic a proof of immortality. Gerhardie's epistemologically agile narrative anticipates the later experiments in phenomenological dislocation in the novels of Muriel Spark.

By implication Gerhardie's novel explores the relevance of the paranormal to a spiritual understanding of the supernatural; and it does so in a tone of cool intelligence and humour rather than of sentiment. The same thing might be said of a novel that came out two years later, *A Harp in Lowndes Square* (1936) by Rachel Ferguson (1893–1957). This book is of particular interest in showing how a supernaturalist element can be accommodated within what would otherwise be the characteristic naturalistic style of a light novelist. Ferguson enjoyed considerable popularity in the 1930s and 1940s. Her speciality was the novel of family life, full of gossipy, confiding humour, lovingly detailed accounts of domestic trivia and an interest in actors and acting; she was an amusing parodist and a regular contributor to *Punch*. Repetitious though her books are, they have a style and an atmosphere all their own; and their author had the rare virtue in a successful writer of subjecting to internal scrutiny the world that made her popular. She does this most tellingly in *Evenfield* (1942), set in a London suburb and describing how a woman's obsession with the world of her childhood leads her to repurchase the family home, and, by refurnishing it as it was, to recreate the past by *force majeure*. The book is a materialistic ghost story, and the tension set up, between the author's fascination with her protagonist's world, and her simultaneous distrust of it, is painful and even creepy. An undercurrent of psychic peril haunts the book.

A Harp in Lowndes Square is a more diffuse handling of a similar theme, but here a preternatural element provides a distancing device that makes for greater balance. Following a brief retrospective prologue, in which a small girl leaning over the banisters of a big house listens to the voices of her future son and daughter, the novel commences with a blithe aplomb suggestive of the style of Rose Macaulay: 'It is on record that when mother found that she was going to have a baby she said to father, "Oh Austen, look what you've done now!"' The child in question is to be followed by identical twins, possessed of total rapport and the gift of second sight: the girl, Vere, is the narrator, and a corollary to her possession of 'the sight' is the knowledge that all time is simultaneous. In this instance such knowledge is an experience of pain. Their mother's suppressed memories of the ill-treatment of her crippled sister at the hands of her own mother, a hard-hearted society woman, overshadows the otherwise idyllically happy childhood of the twins and their elder sister, in whom the spirit of the persecuted Myra is reborn. The novel is steadily prosaic in tone, and the supernaturalist elements are so materially accounted for as to reduce the sense of mystery to a minimum. Vere describes her psychic experience as 'purely photographic, with no free will at all'. Eventually she begins to see a ghost herself, as an intentional manifestation on its part. Here too, however, the stress is on normality, on friendliness and not on terror.

An interesting gloss on the supernatural story is provided when an actor advises Vere to resort to theatricality in order to disperse the oppressive obsession with her mother's past: even the wicked grandmother is not demonic, and can be upstaged.

And if you are ever in danger of making a fetish of your memories come to me and I'll find a set of situations for you far more harrowing than anything *you* ever thought of! And when you go and see Lady Vallant next and find yourself trending towards the old slavery to her manner or expression or words, say to yourself 'Furnival knows an elderly actress who'd do this sort of thing far better'. Artificializing . . . the veil of illusion . . . it's a very healing thing. (ch. 31, author's ellipses)

In such a passage the ludic use of supernaturalist motifs, in order to exhibit human frailty and error, is anticipated to characteristically breezy effect.

If sequential time is partly illusory, and the spiritual world is the real one, then only through artifice can one protect oneself against it: the novel itself is a form of stabilization, as Vere recalls her story. The spiritual presences are mediated through material substance.

That house is *stained* with memories of unhappy things, and those of us who own blood kinship with the movers in that story may be peculiarly susceptible to what it still may do. That is why it is futile to destroy the staircase. Until that picture has worn itself out with time or the superimposition of a set of events serene and normal, and until the atmosphere has so been reconditioned, that picture will remain, ever re-enactable, eternally to be guarded against. (ch. 33, s. ii)

A Harp in Lowndes Square presents a pseudo-scientific comment on the fatalistic novel of the Gothic, demonic kind; but it does so by resort to imaginative concepts which belong in essence to the occultist tradition. More interestingly, Ferguson makes play with different levels and planes of reality for purposes of moral and psychological comment in a way that anticipates such post-war novelists as Iris Murdoch and John Fowles. For all their cosy 'period' feel, her novels are in this respect ahead of their time.

In hermetic fiction the preternatural undergoes an imposition of theory. Because novels of this kind embody an intellectual exploration of space and time and of the secrets behind appearances, and because they are governed by intellectual considerations, they tend to suffer from imaginative preconditioning: the spiritual laws of the mysterium being regarded as open to investigation and ultimate mastery, the origins of life that are symbolized by the figure of the Great Mother merely await their unveiling— as the very title of Madame Blavatsky's theosophical compendium attests. Whereas the tale of terror plays on psychic insecurities, the hermetic tradition grows out of dreams of mastery and power: it belongs to the world of Faust. But it usually distrusts its own subject-matter and, as with Faust, the mastery is paid for dearly.

In the case of the protagonists in Arthur Machen's fiction, the dialectic between materialistic and spiritual concepts that energizes the relations between Frankenstein and his creature, and which is spelt out in moral terms by Bulwer-Lytton and Meade Falkner, becomes swallowed up retributively in preternatural horror. Ironically it is Machen whose work most retains its vitality and persuasiveness. Again, where Blackwood is concerned, didacticism may predominate, but it is when fear erupts and speaks through desolate landscapes and equally desolate urban lodging-houses, which in the best of his stories take on a numinous significance, that his writing attains authentic imaginative power.

Since the study of the occult is practised as a science, no less than in the tale of terror is the reader in a world of laws and prohibitions; and laws can be broken and prohibitions flouted. This is a prescriptive literature rather than a prophetic one. In it the challenge of naturalism is met by plain rebuttal: there is no question of persuasion or reciprocal interpretation, still less of irony. Of the hermetic writers discussed here, only Machen and Blackwood move from enquiry as to the riddle of the universe to an authentically personal celebration of its mystery. In their fictions terror is replaced as a governing principle by awe, and they are at their most convincing when they attempt to demonstrate the inescapable connection between physical laws and the ones they propose as governing the movements of the human spirit.

4

AN INSINUATION OF
DOCTRINE

No one could ever have found God; he gave himself away.

(Meister Eckhart)

In a striking passage from *Memories, Dreams, Reflections* (1961)
C. G. Jung describes the sense of incongruity felt by him as a boy
when considering the supreme mystery of what the Christian
Church exists to proclaim, and the devaluing nature of the proclamation itself:

people were exhorted to have those feelings and to *believe* that secret
which I *knew* to be the deepest, innermost certainty, a certainty not to
be betrayed by a single word. I could only conclude that apparently no
one knew about this secret, not even the parson, for otherwise no one
would have dared to expose the mystery of God in public and to profane
these inexpressible feelings with stale sentimentalities. (ch. 2)

The containing of the Church's treasure within earthen vessels is
a problem that faces any writer wanting to present a fresh and
worthy articulation of his or her religious belief. The mysterium
eludes embodiment, withdrawing in the face of definition. Such
an impasse provides a challenge to authors of supernaturalist
fiction: the literary methodology they employ allows for an alternative vocabulary to the customary hackneyed one, and with
linguistic changes there comes a reillumination of obscured truths
and teachings. The disparity that Jung deplored does, however,
go deep: what applies to the normal verbal usages applies likewise to the literary didact (whose own nature enshrines a similar
contradiction). The apprehension of religious truth can only be

attained obliquely, and in most cases by a deliberate side-stepping of the official voices and dispensers of the mystery in favour of the voice of inner personal knowledge and experience.

In *Wuthering Heights*, for example, apart from the Calvinistic mutterings of old Joseph, traditional Christianity plays virtually no role at all. Despite her upbringing in a Church of England parsonage, Emily Brontë eschews ecclesiastical religion: her mystical temperament had no need of it. Her sister Charlotte, on the other hand, while sharply critical of the clerical establishment, does portray its ministers in their pastoral and spiritual capacities (although the portraits she provides are scarcely flattering). *Jane Eyre*, indeed, has some claim to be regarded as a supernaturalist novel, and evokes a complex interplay of other-worldly elements that in turn interact, not to say conflict, with the conventions of the version of Christianity to which its author formally subscribed. Charlotte Brontë's peculiar imaginative gifts so shape the element of narrative mystery that it spills over from the intricacies of plot into an evaluation of plot's content. Jane Eyre's story moves from an illusory experience of the preternatural in the red-room at Gateshead Hall, through Gothic terror in the attics at Thornfield, into the realm of providential religious mystery. The latter stage occurs when Jane, about to commit herself to the tendentiously named missionary St John Rivers, is exposed to the miracle of transcendent grace through the blinded Rochester's telepathic summons. The plot evolves from the preternatural into the supernatural; it parallels and comments upon Jane's inward journey from terror and resentment, through dread and dedication, to awe and the fullness of love-in-mystery. Instead of providing a supernaturalist metaphor for a natural process, *Jane Eyre* is a naturalist metaphor for a supernatural one.

Such a conflation was rare in an age when the establishment of Christianity on a formal basis as the national religion meant that the custodians of that religion were implicated in the structures of the society that sustained it; for whatever may have been the case with the beliefs of their readers, the refusal of automatic assent to orthodox theology by the most esteemed late nineteenth-century novelists (George Eliot, Meredith, Hardy, James) meant that Christianity itself became a specific subject for fictive treatment. Writers such as 'Mark Rutherford' and Mrs Humphrey Ward, who dealt specifically with religious doubt, had fewer literary problems than had those who wrote of religion in a spirit of belief: how to avoid producing what was in effect a tract? *Loss and Gain* (1848), Newman's account of the conversion of a young Anglican to Rome, shows that in this respect keen intelligence and theological subtlety are not enough to secure an engaged emotional interest, especially when exercised within such a narrowly ecclesiastical focus. The historical method, as deployed in J. H. Shorthouse's popular *John Inglesant* (1880), with its portrayal of seventeenth-century Anglican and Roman Catholic piety, provides a more readily assimilable medium (H. F. M. Prescott's *The Man on a Donkey* (1952), an account of the Pilgrimage of Grace, provides a moving and impressive twentieth-century instance); but for novelists concerned with contemporary religious experience, the very existence of a stratified clerical hierarchy stood in the way of the imaginative depiction of the supernatural realities they existed to maintain. The truths and speculations concerning religious experience were obscured by a concern with its ecclesiastical embodiments.

George MacDonald

It is therefore not surprising that by the mid-nineteenth century an abundant crop of not very impressive clergymen had come to flower in the pastures of the English novel. Goldsmith's Dr Primrose is an exception in his simple goodness, and though Parson Adams may lend an element of innocence to the tough realities of Fielding's England, Thwackum and Trulliber provide a crushing counterweight. The worldliness of Peacock's clergymen is more polished, but worldly none the less; while Jane Austen's Mr Collins, Dr Grant, and Mr Elton are objects of mockery or, at best, dispassionate satire. And Dickens is unsparing of the tippling Stiggins and the oily Chadband, both of whom are portrayed as predators on a congregation of gullible subscribers. Although Trollope may exercise a more benevolent patronage in *The Warden* (1855), in *Barchester Towers* (1857) he waxes disrespectfully merry at the expense of the Anglican episcopal establishment; while Charlotte Brontë is caustic about North Country curates and Thackeray over West End preachers. In not one of these cases do the clergy suggest the priesthood of a transcendent realm of supernatural reality. Only *Scenes of Clerical Life* (1858) offers an exception, and ironically George Eliot was agnostic, though her Evangelical upbringing had acquainted her with that experience of a spiritual order which is at the heart of genuine religious belief. But it is surely significant that the clergy should have been hounded with such persistence by the more thoughtful talents of the day: it is as though the disparity between their involvement in the society those novelists either despised or seriously questioned, and their custodianship of a metaphysical mystery which implicitly contradicted that involvement, proved too outrageous a paradox to be condoned.

The corrupting nature of the discrepancy is made tellingly obvious at the opening of *Jane Eyre*. Charlotte Brontë here provides a classic instance of the terrors that can be engendered by a guilty rage aroused by the oppressive perversion of the religious spirit in guardians and mentors. The scene opens with an independent-spirited child shut up in a lonely bedroom in which the master of the house has died, a place of massive looming furniture, of sadness and terrifying cold. In a marvellously imaginative passage Brontë portrays the generating of preternatural dread and the orphan's sense of being rejected and thus of being herself unnatural.

Daylight began to forsake the red-room; it was past four o'clock, and the beclouded afternoon was tending to drear twilight. I heard the rain still beating continuously on the staircase window, and the wind howling in the grove behind the hall; I grew by degrees cold as a stone, and then my courage sank. My habitual mood of humiliation, self-doubt, forlorn depression, fell damp on the embers of my decaying ire. All said I was wicked, and perhaps I might be so: what thought had I been but just conceiving of starving myself to death? That certainly was a crime: and was I fit to die? Or was the vault under the chancel of Gateshead Church an inviting bourne? In such a vault I had been told did Mr Reed lie buried; and led by this thought to recall his idea, I dwelt on it with gathering dread.

The logical bent of Brontë's imagination lends a further turn of the screw, which is itself a critique of the attitudes which have consigned the child to such a punishment.

I doubted not—never doubted—that if Mr Reed had been alive he would have treated me kindly; and now, as I sat looking at the white bed and overshadowed walls—occasionally also turning a fascinated eye towards the dimly gleaming mirror—I began to recall what I had heard of dead men, troubled in their graves by the violation of their last wishes, revisiting the earth to punish the perjured and avenge the oppressed; and I thought Mr Reed's spirit, harassed by the wrongs of his sister's child, might quit its abode—whether in the church vault, or in the unknown

world of the departed—and rise before me in this chamber. I wiped my
tears and hushed my sobs, fearful lest any sign of violent grief might
waken a preternatural voice to comfort me, or elicit from the gloom
some haloed face, bending over me with strange pity. (ch. 2)

That haloed face, a blend of chastisement and love, is one that
was familiar in the Victorian iconography of guilt. The presence
of a spiritual overworld of messengers and guardians mirrors the
essentially parental nature of nineteenth-century notions of Divine Providence; and with it there goes the belief that Heaven has
to be earned—it is the reward of moral enterprise, just as earthly
wealth is the fruit of business enterprise. This essentially Pelagian
interpretation of the efficacy of good works is, however, modified
by the Evangelical emphasis on grace, on the free bestowal of
God's love as being both the source and nature of His creation.
If, through their socially exemplary role, the clergy inevitably
emphasized the element of works, their secret ministry to individuals was more subversive of accepted standards. A conflict between the two emphases could arise explicitly in the self-governing
bodies of Dissent, when a minister's theology became too accom-
modating for the self-enclosed selectiveness of individual congre-
gations to stomach. A notable instance of such a collision of
beliefs lies behind the abandonment by George MacDonald (1824–
1905) of his ministry to his Arundel congregation, a resignation
forced upon him by his preaching too liberal a gospel of salvation.

MacDonald, indeed, is arguably the most spiritually minded of
nineteenth-century Christian novelists, possessing as he did 'a
strong conviction that a transcendent reality is so closely related
to the world of immediate human experience that it is also immanent in it'.[1] His exclusion from the sacred ministry allowed
him freely to articulate his theological beliefs in more than twenty
novels and in a number of fantasies and tales for children. The
simple distinction of genre cannot do justice to his profound
sacramentalism: his novels, most of them naturalistic studies of
Scottish provincial life, are filled with a latent symbolism, suggestive of a spiritual order watching over and protecting the
fortunes of his characters. The everyday is shot through with the
numinous, as in the strange elfin-like protagonist of *Sir Gibbie*
(1879) or in the various legends and fairy tales that punctuate
the narrative of *Adela Cathcart* (1864). Where it is more overtly
supernaturalist, his writing is less successful: an early work,
The Portent (1860), designated 'a Romance', is ostensibly a tale

concerning the second sight; but its insistence on ancestral doom, on the extrasensory perception of a spae-wife and the telepathic affinities of the lovers, are not linked in any coherent way, and are relegated to the domain of the merely marvellous, being dismissed at the end in the name of a higher, religious reality.

It was as if the gates of the unseen world were closing against us, because we had shut ourselves up in the world of the present. But we let it go gladly. We felt that love was the gate to an unseen world infinitely beyond that region of the psychological in which we had hitherto moved. (ch. 27)

Although this is an orthodox conclusion from a Christian point of view, the preceding events have left one ill-prepared for it.

More characteristic of MacDonald's particular vision is the children's story *At the Back of the North Wind* (1871). Whereas *The Portent* uses the threadbare properties of Romanticism, this book combines a didactic strain with an occasionally sharp awareness of social conditions in the London slums. A small boy called Diamond encounters North Wind, a beautiful woman who can change her size at will; she shows him the workings of the creative energy of God, preaching an affirmative wisdom and finally being with the boy at the hour of his death. She is a mother-figure, a protectress, and an instrument of redeeming providence. She is also the gateway of death: to reach the heavenly country at the back of the north wind Diamond has to go *through* the wind. The integration of natural with supernatural is perfectly achieved in MacDonald's account of the boy's experiences: his dream vision of his adventures with North Wind are his personal version of the condition of sleepwalking and subsequent illness; his physical and spiritual states coalesce, and death is seen as a dimension of life—a central theme in the author's credo. Diamond's own death is genuinely pathetic because it is the natural fulfilment of what has gone before. MacDonald shares the nineteenth century's preoccupation with child mortality, but his treatment of it is rarely mawkish, because he presents the subject less with an eye to pathos than as evidence of his own deeply held convictions concerning the reality of a supernatural world.

The sense of that world impinges strongly on his two finest books for children, *The Princess and the Goblin* (1872) and *The Princess and Curdie* (1882). Though classed as fairy-tales, they may also be read as psycho-dramas, as narratives that deal with the properties and spiritual laws of the human soul, 'set in land-scapes which are symbols of mind and are concerned with men-tal perception'.[2] They also dramatize the conflict between the spiritual vision of the imagination, and the self-centred obfuscating forces of materialistic greed; and they Christianize it, endowing romance with moral purpose.

This is nowhere more apparent than in MacDonald's supreme imaginative creation, the young Princess Irene's great-great-grandmother, a wondrous figure, infinitely old, yet young and infinitely beautiful and strong, who lives above the deserted up-per storeys of the child's mountain home. MacDonald is no pre-cise allegorist as is Bunyan, whom in some other aspects he resembles; and the old Princess Irene is an entirely autonomous creation, sharing certain characteristics with other of his bene-volent all-powerful female figures, such as North Wind and the eponymous protagonist of *The Wise Woman* (1877), who pos-sessed 'the old age of everlasting youth' (ch. 13). Indeed, in these figures MacDonald shows himself a theological innovator, pre-senting a convincing image of the femininity both of Godhead and of the operations of divine grace. But the old princess also embodies the power and purpose of romantic vision, the know-ledge of the spiritual order which is the heart of the material one. It is she who provides her granddaughter with the woven thread that guides her safely through the darkness of the mines; whose flock of pigeons act as messengers of grace that give birth to in-tuitive perceptions; and whose fire of roses purges and strengthens those who are bathed in its terrible beauty. She also gives to the young miner Curdie the power to recognize the true nature of any human being whose hand he clasps, endowing those whom she calls to serve her with spiritual vision of total clarity.

The old princess presides over these two otherwise rather dif-ferent stories and gives them unity. Indeed, one might argue that it is she and not the young Irene who is the princess of both titles. For who is 'the goblin' of the earlier book? None of that corrupt underground brood has any particular relation to the little girl: only the plural generic designation would meet that particular case. If the title be read as allegorical, then the self-seeking, materialistic, blinded goblin in humankind is contrasted with the eternal life of the visionary spirit, whose healing role the old princess enacts. And in the sequel, which seems to be aimed at an older readership, she empowers Curdie, incarnation of loyalty, truthfulness, and practicality, to overcome the corruption that has imprisoned the young Irene and her father, the king. As much a social fable as a fairy tale, *The Princess and Curdie* dramatizes the operations of the visionary spirit in a materialistic world.

MacDonald's most original, though not most satisfactory, achievement is *Lilith* (1895), the second of the two fantasy narratives which frame his literary career. The imagery of *Phantastes* (1858) is strongly influenced by German Romanticism and the writings of Novalis and La Motte Fouqué. It is pure fantasy, a narrative that propels itself through something approaching free association; but *Lilith* relates more closely to the conventions of the traditional novel. It may be designated a mythopoeic psycho-drama, a narrative dealing with archetypal figures (in this instance Adam, Eve, Lilith) that relate to the soul's awakening to self-knowledge and repentance into eternal life. The book is a portrait of the spiritual universe as perceived by one voyager within it. What the narrator is doing he is at the same time being: to this extent *Lilith* resembles *The Pilgrim's Progress*. The material and spiritual worlds are portrayed as 'coincident and co-existent'.

Part of the success of *Lilith* comes from its mastery of dream-narrative. As W. H. Auden comments, 'the illusion of participating in a real dream is perfect; one never feels that it is an allegorical presentation of wakeful conscious processes'.[3] *Lilith* is no simple allegory; but its form has some bearing on the relation between a naturalistic novel and any supernaturalist content its author may seek to include within it. The narrator, Mr Vane, walks into the world of his dream through an interplay of light between two mirrors in his attic. Whenever he does something in the dream world that breaks the law of his presence there, he finds that he has returned to his house and garden, set back in the more constricted world of 'the three dimensions'. In literary terms this reads like a parable concerning the misunderstanding of the supernatural which arises from a confusion of categories or a misappropriation of one of them in the service of another. Whatever else, MacDonald through these means indicates that the world of matter is enclosed in a metaphysical universe with its own inexorable spiritual laws: the dramatic devices, both here and in his books for children, make such a view entirely plausible.

Still more notable is the shining optimism of his outlook: the metaphysical world, although it has its terrors (physically evident in *Lilith* with its monsters of the Bad Burrow and the vampiric proclivities of Lilith herself) is suffused with a sense of redemptive mercy and grace. Not that there is any sentimentality about MacDonald's presentation of these things: his awareness of how people shut themselves off from mercy is reflected in the account of Lilith's knowledge of her own life-in-death. 'She knew life only to know that it was dead, and that, in her, death lived. It was not merely that life had ceased in her, but that she was consciously a dead thing. She had killed her life, and she was dead—and knew it.' This analysis of the damnation of despair is only made more incisive by the narrator's previous comment that

We were not in the outer darkness; had we been, we could not have been *with* her; we should have been timelessly, spacelessly, absolutely apart. The darkness knows neither the light nor itself; only the light knows itself and the darkness also. None but God hates evil and understands it. (ch. 39)

MacDonald's theological insights are frequently embodied in images that enact the spiritual processes they signify. The simple language of the 'children', the simple in heart who have to grow up if they are to overcome the deathly power of Lilith, mistress of illusion, captures graphically the working of the Shadow that accompanies and enslaves her.

He was nothing but blackness . . . He came on us as if he would walk over us. But before he reached us, he began to spread and spread, and grew bigger and bigger, till at last he was so big that he went out of our sight, and we saw him no more, then he was upon us! . . . He was all black through between us, and we could not see one another; and then he was inside us. (ch. 37)

Such forceful writing justifies the claim of his most recent biographer that for MacDonald 'Faerie is the real world, while this one is but a pale shadow of the other.'[4] Like Mr Vane, he could claim that

I was constantly seeing, and on the outlook to see, strange analogies, not only between the facts of different sciences of the same order, or between physical and metaphysical facts, but between physical hypotheses and suggestions glimmering out of the metaphysical dreams in which I was in the habit of falling. (ch. 1)

It is not surprising that MacDonald should be regarded as the forerunner of later Christian fabulists and fantasists; C. S. Lewis, J. R. R. Tolkien, and Charles Williams owe much to him. However clumsy or stumbling his prose can be, however morbid or archly sentimental in mid-Victorian fashion (he lacks the convincing grossness of Le Fanu), he is a genuine visionary for whom the world itself is evidence of things not seen. At his finest, in much of *Phantastes*, in the character of the old princess in *The Princess and the Goblin* and its sequel, and in the sombre closing pages of *Lilith*, he persuades one not only of the beauty of holiness as an ideal (currently a most unfashionable concept) but also, with an appeal to self-knowledge, as a reality that involves terror as well as joy and comfort. Such optimism and assurance as to the benevolence of God is on a different level from the sentimental meliorism of the piety of his time. In MacDonald's work the Gothic and the preternatural are baptized: it embodies an altogether loftier imaginative world.

A kindred experience can be undergone in reading the ghostly tales of his contemporary, Margaret Oliphant (1828–97). *A Beleaguered City* (1880) and *Stories of the Seen and Unseen* (1902) are notable for their religious, at times theological, content. Their author had herself suffered much personal bereavement at the time she wrote these tales, many of which depict communication between the living and the dead. Her ghosts are of real people, not spectres or goblins; and their concerns are thus in the strictest sense supernatural. She can produce some fine preternaturalist effects, such as the unearthly weeping in her most anthologized story, 'The Open Door'; but, more frequently, strangeness is subsumed into the whys and wherefores of the apparitions' presence; in some cases, indeed, the visitor from beyond the veil is apparently as 'real' as are those on this side. In 'Earthbound', the charming revenant is doing penance for her refusal to surrender the delights of bodily existence; while in 'Old Lady Mary', a kindly but self-indulgent woman is allowed back on earth in order to put right, if she can, a negligence concerning the disposal of her property. The originality of this particular story lies in its being told both from the point of view of those who hear the ghost and thus experience her as preternatural, and from that of the ghost herself. Another unusual feature is that, where putting her affairs in order is concerned, her quest fails; but matters do get put to rights through a seeming accident and by the natural course of events, the real fulfilment of her mission being the exchange of love and forgiveness across the grave with the girl she has unintentionally injured.

Ironies of this kind are likewise present in the short novel, *A Beleaguered City*. The Burgundian town of Semur, having fallen into godless ways, is suddenly enveloped in darkness and occupied by the spirits of the righteous dead who, with silent, invisible compulsion, drive the inhabitants outside the gates. After a while the blessed ones depart, the light returns, and the citizens reoccupy their homes. The story is told by a number of witnesses, chief of whom is the Mayor, a well-meaning but complacent and materialistic man, who believes in religion as a socially cohesive force, rather than as the expression of a supernatural reality. The story is confusing not least because the reason for the visitation is unclear: is it or is it not the discontinuation of a Mass said at the city hospital? A further irony is that the repentance of the citizens is short-lived: everything is soon exactly as it was before. However feeble as drama, the tale does carry conviction as a parable. As with all Oliphant's stories of this kind, it conveys a genuine sense of a supernatural dimension encompassing everyday affairs. In other stories, such as 'The Library Window', material reality is evoked so vividly that small physical details actually enhance the supernatural effect. As with other Victorian writers engaged with this theme, Margaret Oliphant's success was partly based on her response to physical environment. All the finest stories in this tradition have a firmly naturalistic base.

Four Christian Apologists

The genre of fantasy cannot be held a genuine part of the supernaturalist tradition: its relation to it is implicit or indirect. But it is frequently used as a short-cut to the presentation of supernatural reality. MacDonald's deployment of it to project a Christian message was adopted in the early twentieth century by G. K. Chesterton (1872–1936), C. S. Lewis (1898–1963), and J. R. R. Tolkien (1892–1973). The last-named's Middle Earth, while it is full of supernatural powers and forces, is, save for its origins in traditional fairy tale and myth, a world entirely removed from this one. It embodies an understanding of life that is magical and spiritual, in the Blakean sense; the supernatural operates within it administratively rather than organically, as it does in saga. Lewis's tales of Narnia (written for children and published between 1950 and 1956), although more overtly theological in tone than Tolkien's *The Lord of the Rings* (1954–5), belong to the same tradition.

The free flow of association that marks MacDonald's super-naturalist fiction is less evident in his successors: because they are avowed propagandists, Chesterton and Lewis tend to be hampered by prescriptive affirmations. On the other hand, an insistence on paradox and on the alienating nature of religious belief, such as one finds in the early novels of Graham Greene, present complementary difficulties as to the command of imaginative assent.

Where Lewis is concerned, one is confronted with an essentially derivative and bookish imagination, nourished on scholarship and extensive reading within selective fields. His most personal, subtle, and satisfying novel, *Till We Have Faces* (1956), is a retelling of the myth of Cupid and Psyche from the point of view of the latter's jealous, ugly, and self-resentful sister Orual, whose tormented rejection of Psyche's vision and knowledge of her heavenly lover enacts the rationalistic mind's complex attitude towards the evidence for supernatural experience itself. Lewis here aligns the paradox inherent in romantic vision with the intellectual demands posed by theodicy, the attempt to reconcile God's goodness with his power. Orual's possessiveness blinds her to the world that Psyche knows; but she also rejects the possibility that the vision may be true, denying her own momentary glimpse of the palace of love. In this book Lewis overcomes his usual tendency to impose a didactic morality upon a mythological story; and he does so by resorting to psycho-drama in the manner of his revered MacDonald. At the novel's end Orual realizes that she is herself Psyche: the retelling of an ancient legend has become the exposition of a permanent spiritual reality, an interior myth replacing a ritual one.

Lewis's one exercise in the genuine supernaturalist tale is *That Hideous Strength* (1945), its predecessors, *Out of the Silent Planet* (1938) and *Perelandra* (1943) belonging more to the genres of fantasy and science fiction. They are, however, his most original imaginative creations, revealing his gift for embodying metaphysical concepts in unusual forms, the portrayal of outer space in terms of light and colour reflecting his knowledgeable assimilation of the medieval world-view so eloquently set forth in his study, *The Discarded Image* (1964). Equally characteristic is the account of the planetary angels in *That Hideous Strength*. An observer

would have known sensuously, until his outraged senses forsook him, that the visitants in that room were in it not because they were at rest but because they glanced and wheeled through the packed reality of heaven (which men call empty space) to keep their beams upon this spot of the moving earth's hide. (ch. 15)

This kind of theological conceit is more persuasive than are the author's frequent moralizing interjections.

In this novel Lewis makes use of contrasting literary tactics to demonstrate that any attempt by a power-hungry scientific conglomerate to draw on cosmic spiritual energies is bound to fail, since it does not possess the kind of understanding that can safely treat with them: the materialistically minded are objects of scathing, all-too-knowing satire. But as an imaginative counter-blast to the ruthless world of scientific and university politics the private, almost nursery world of St Anne's-on-the-Hill, with its wounded Mr Fisher-King, comic servant and tame bear, is quite inadequate (much as the snug domestic world of Sol Gills and Captain Cuttle is an inadequate moral counterweight to the cold world of Victorian materialism portrayed by Dickens in *Dombey and Son*). There is a sense of some privileged, private access to the truth which is the very obverse of Psyche's knowledge of her lover in *Till We Have Faces*—but the possibility and incidence of which in much religious fiction to some extent justifies Orual's passionate rejection of it.

In its ideas and its deployment of preternaturalist effects, *That Hideous Strength* shows the influence of Charles Williams; but it lacks that fusion of spiritual and material categories which in Williams's novels serves logically, and thus convincingly, to resolve the plot. For all its great readability and cleverness, the didactic spirit in which it is written and the author's role as ringmaster deprive the book of any real sense of the numinous. None the less, its frequent moments of beauty, insight, and humour are the outcome of a genuinely supernaturalist perspective. Its controlling theological vision is based on popular history and a sharp awareness of what may be called the romantic conscience: Lewis's belief in the validity of informed aesthetic response is nowhere more evident than in his account of Bragdon Wood, the burial place of his earthy, vigorous, pagan Merlin. Here he is at his most engagingly spontaneous and enthusiastic, evoking the kind of English folklore found in Kipling's *Puck of Pook's Hill* (1906), in which oral tradition gives rise to a mythology born of a shared and worked, and thus familiar, landscape.

Englishry is likewise pervasive in the writings of Chesterton, not least in his novels, if novel be the word for works with such cheerfully flagrant designs upon their readers. Stories like *The Man who was Thursday* (1908) are hard to classify, though easy to enjoy. Theologically this is the most interesting of the novels. Six members of the Central Anarchist Council, named after the days of the week, having discovered that they are all policemen in disguise, set out in pursuit of Sunday, the master anarchist, only to discover that he is, to all intents and purposes, God—a buoyant, jesting God, at once frivolous and awe-inspiring, and quintessentially Chestertonian in his optimism and mastery of paradox. Although the tale takes place in the contemporary world, it cannot, despite its portrayal of the theological nature of the supernatural, be regarded as a serious supernaturalist novel: it belongs more in the class of the fantastic fable, with stories like Stevenson's *New Arabian Nights* (1882) or *The Green Overcoat* (1912), Hilaire Belloc's balefully debonair satire upon spiritualism (and much else). At one point it reads like a corrective to the devolutionary extravagances of Arthur Machen. Tuesday, the pessimist, is recounting his first meeting with the master anarchist.

He sat there on a bench, a huge heap of a man, dark and out of shape. He listened to all my words without speaking or even stirring. I poured out my most passionate appeals, and asked my most eloquent questions. Then, after a long silence, the Thing began to shake, and I thought it was shaken by some secret malady. It shook like a loathsome and living jelly. It reminded me of everything I had ever read about the base bodies that are the origin of life—the deep sea lumps and protoplasm. It seemed like the final form of matter, the most shapeless and the most shameful . . . And then it broke upon me that the bestial mountain was shaking with a lonely laughter, and the laughter was at me. (ch. 14)

The reversal is itself a jest at the expense of *fin de siècle* occultism.

Laughter is essential to Chesterton's outlook: his hatred of contemporary materialism is partly activated by its inherent (as he understood it) gloom, leading him, for instance, to the notorious designation of Hardy as 'a sort of village atheist brooding and blaspheming over the village idiot'.[5] For Chesterton, the world itself is supernatural, and hope not simply a virtue but, as one of the theological virtues, an attitude of moral accuracy: it is based on fact. None the less he is acutely conscious of the demands of theodicy. As Thursday remarks, 'Bad is so bad, that we cannot but think good an accident; good is so good, that we feel certain that evil could be explained' (ch. 14). For this writer, the really damning thing is the refusal to think or to acknowledge that the paradox needs to be faced, and questions to be asked. He sees the ultimate betrayal as being the acquiescence in a total relativism, 'that final scepticism which can find no floor to the universe' (ch. 11). But the knowledge of the mysterium will always be a knowledge of uncertainty. It is a knowledge from which Chesterton's temperament shied away, which may be why, for all the gaiety and originality of his imagination, his work never quite attains to the transfiguring uncertainties of a revelation of the numinous. His is a declaratory rather than a persuasive art. Nevertheless, its basic premisses are those on which any supernaturalist literature must build if it is to avoid the limitations of merely diversionary entertainment.

Although, like Lewis, Chesterton is a didact, his incarnational theology springs more resoundingly from the heart. His fiction voices a belief that the natural is itself mysterious; he writes fantastic fables in which the extraordinary nature of the ordinary is affirmed as paradox. His Father Brown stories suggest an allegorical form of detection which penetrates the world of everyday to demonstrate the existence of the spiritual universe encompassing it. But the shifting perspectives, the dislocating awareness of the twentieth-century materialistic consciousness are lacking. For better or worse, Chesterton's work proceeds from certainties, and is to that extent self-limited, being rooted in late nineteenth-century conventions and beliefs, not in twentieth-century relativism.

It possesses, however, a light-heartedness which is infectious: if it fails to evoke the numinous, it none the less points to a state of being which exposes even human spiritual apprehensions as necessarily relative. In this respect Chesterton's work contrasts happily with the polemical novels of Robert Hugh Benson (1871–1914), whose books included Catholic historical propaganda such as the gustily entitled *Come Rack! Come Rope!* (1912), theologically motivated studies of contemporary life (*Initiation* (1914) contains spiritual teachings also found in the later novels of Charles Williams), and apocalyptic extravaganzas like *The Dawn of All* (1911). *The Necromancers* (1909) takes a critical look at spiritualism and black magic from the pseudo-scientific perspective of writers in the hermetic tradition; but far more persuasive as supernaturalist fiction than all these experiments in diverse modes are two collections of ghost stories, *The Light Invisible* (1903) and *A Mirror of Shalott* (1907), which, by focusing on the pastoral experiences of Catholic priests, subdue the genre to a controlling theological pattern, again in the manner of Charles Williams. Both writers portray occult studies as dangerous, and Christian supernaturalism as affirmative and sane. In this they reflect the influence of fictive naturalism, which of its very nature has to posit that the material world is one in whose significance it is necessary to believe. Such a conviction is likewise the basis of Christian sacramentalism, in which body and spirit are mutually interdependent. But Benson's fictions have a dramatic crudity that renders them imaginatively dualistic. The divine sovereignty is compromised: as Williams was subsequently to remark in another connection, the pious always are 'in a state of high anxiety to defend and protect, and generally stand up for, Almighty God'.[6]

That observation has some relevance to the fictions of another and later Catholic novelist, Graham Greene (1904–91)—at any rate to those in which his religious allegiance makes itself directly felt. The dualism they contain forms part of an innate pessimism and of a perception of the implications of divine transcendence. *Brighton Rock* (1938), with its evil but reluctantly believing Catholic boy-gangster, Pinkie, is a theological fable; but it also teases out a problem of imaginative epistemology—how meaningfully to employ terms that have lost their original intellectual and emotional context? Can the categories of good and evil be usefully employed by a twentieth-century writer of fiction? Or must they be regarded as synonymous with right and wrong, differing only in their superior intensity? The very understanding of the term 'supernatural' is bound up with that question.

Greene's engagement with contemporary disbelief in the supernatural is at its most complex and unresolved in *The End of the Affair* (1951). The narrator is himself a novelist, whose involvement with the wife of a civil servant has come to an end eighteen months before. The book describes his realization that he has not lost her to another man but to God: in the words of her diary, 'I've caught belief like a disease. I've fallen into belief like I fell in love' (book v, ch. 1). Her ex-lover's furious rejection of this conversion, his refusal to accept her belief, leads to her death, and thus to her subsequent performance of two apparent miracles. His self-hatred is expressed with the painful force of which Greene is such a master, and in itself conditions one towards accepting the metaphysical implications of the story. His very hatred of the idea of God implies a belief in his existence.

Greene does not merely permit the miraculous element to speak for itself. As so often, he allows a Catholic priest to voice a delimiting comment: 'I'm not against a bit of superstition. It gives people the idea that this world's not everything . . . It could be the beginning of wisdom' (book v, ch. 7).[7] Yet the healings associated with the dead woman's 'relics' (a children's book, a lock of her hair) are real enough: cause and effect are not absolutes as such, being rooted in the awareness of chronological time. But, as the priest observes, 'St Augustine was asked where time came from. He said it came out of the future which didn't exist yet, into the present that had no duration, and went into the past which had ceased to exist. I don't know that we can understand time any better than a child' (book v, ch. 7). As the child experiences it, time resembles an eternal present.

The narrator's own conclusion, as a writer, is that 'my realism has been at fault all these years, for nothing in life now ever seems to end' (book v, ch. 1). Greene himself, the most economical and tightly structuring of novelists, subverts his own methodology by the openness of his message. But the subversion is reciprocal; in *The End of the Affair*, as in *Brighton Rock*, the relentless ironies of plot and presentation are self-conscious, for an imperfectly concealed doctrinal imperative is always less disarming than is an avowed intent. Greene's protagonists are at once enmeshed in their beliefs and resentful of them. His portrayal of the world as totally depraved means that in his cosmos the supernatural is experienced as preternatural: since God's ways are necessarily hostile in a sinful world, his love chastises to the death, his presence being known only in a void and an aching sense of loss.

Charles Williams

The extremes of assurance, whether presented openly and cheerfully as in the works of Benson, Chesterton, and Lewis, or with the tortuous reluctance of the more sophisticated Greene, are themselves subjected to questioning by a writer whose attitude to religious faith was altogether more complex and many-sided. The novels of Charles Williams (1886–1945) provide a vision of the physical world as the arena of a sustaining and redemptive Divine Providence; but they draw upon a wide range of supernaturalist traditions and are laced with scepticism and self-doubt. There is nothing polemical about them. Williams himself moved from a committed involvement with the occult (like Arthur Machen and Algernon Blackwood, he belonged to A. E. Waite's Fellowship of the Rosy Cross, a Christian offshoot of the Order of the Golden Dawn) to being a respected lay theologian in the Church of England; and his work deals with the supernatural from a number of viewpoints. In the variety of their modes and forms his novels exemplify much of what was being attempted in the hermetic tradition, while in themselves conducting an examination of the premises that lay behind it. They are essentially theological in inspiration—which is why they are discussed in that context and not in an occultist one. *War in Heaven* (1930), the first to be published, goes directly to the heart of the problem underlying the genre as such.

It adopts plot motifs and imaginative trappings such as one finds in the novels of R. H. Benson, with a dash of Chesterton as well. The discovery of the Holy Grail in an English village church repeats an idea found in Evelyn Underhill's novel, *The Column of Dust* (1909)[7] and in Machen's *The Secret Glory*; other elements, including the frequently flippant dialogue, are found in the detective stories of Williams's friend, Dorothy L. Sayers. The accounts of occult happenings reflect his Rosicrucian studies, and compare interestingly with the specific, almost clinical details found in Huysman's *Là-Bas*, for early twentieth-century novelists a classic text concerning Satanism. Williams's descriptions are meticulous, exact, and less emotively presented than is usual in such accounts: they eschew the kind of rhetoric employed by Benson. Williams describes the turning a man to dust when he steps inside a ritual magic circle; there is an account of the celebration of some kind of Black Mass, and of a solitary partaking of the witches' sabbath through a species of ritual anointing. Despite all their potential blood-curdling dreadfulness, these things are systematically downgraded, presented as petty, infantile, obscene.

As well as being a tale of black magic and a detective story, *War in Heaven* is a work of theological debate concerning the philosophical validity of dualism. Are good and evil commensurate powers? Williams's answer is a resounding No, his story declaring that a provident, inescapable reality is in charge of, and operates through, everything that happens, and not least through the scepticism, so to call it, of the central figure. The Archdeacon of Castra Parvulorum makes short work of those who try to protect the Grail.

'To insult God—' the Duke began.

'How can you insult God?' the Archdeacon asked. 'About as much as you can pull His nose.'

At the same time Williams can scrutinize even his most cherished tenets. When the Archdeacon urges his companions to pray against the diabolic forces unleashed against the Grail, 'It crossed Kenneth's mind, as he sank to his knees, that if God could not be insulted, neither could He be defied, nor in that case the procession and retrogression of the universe disturbed by the subject motion of its atoms' (ch. 10). Williams always insists on the utter transcendence of the absolute.

At the dramatic level, however, there are some philosophical inconsistencies, as in the perplexing appearance on the scene of Prester John. He walks in as a rather dandified young man in a grey suit, who to many of the characters is 'foreign looking'. He elicits and exaggerates the essential nature of the people he meets; he is a touchstone of the truth. But he is also 'the Graal and the keeper of the Graal', and in the final chapter celebrates the Mass and disappears, taking the Grail with him. He is clearly a type of Christ, almost at times Christ himself (Williams more than once in his fiction draws the narrowest of distinctions between the Messiah and his disciples); yet in narrational terms he seems more like an intruding rescuer than a controlling providence. This is a rare example in Williams's work of that reifying of the spiritual which encourages idolatry.

Such uncertainty is characteristic of *War in Heaven*. Williams's theological vision is at odds with his literary material: its roots being in a sense of synchronicity, it does not lend itself to dramatization in terms of chronological duration. The book has wit, sharp insight, a lively narrative flow, a persistent tone of sardonic geniality which makes it among the most readable novels of its kind; but that portayal of the eternal world which was Williams's supreme imaginative endowment is here compromised by dualism. The black magic is all too credible; as a result the power of goodness tends to be invoked as counter-activity, not as an encompassing one—witness the appearance of Prester John. Williams's subsequent novels can be read as so many attempts to offset this confusion.

In each of the next three a controlling image or myth draws the characters into its own sphere of being. Williams writes of such happenings with great power, portraying the physically preternatural in terms of the mental state in which one would observe it: the narrative tone is watchful, informed, absorbed. *Many Dimensions* (1931) is particularly impressive as it deploys the metaphysical possibilities of the Ring of Solomon, which enable its wearers to move in time and space; what for most writers would be a wonder-working piece of fantasy here becomes a parable of the relation of human freedom to Divine Providence. In *The Greater Trumps* (1932) it is the metaphysical implications of the symbols in the Tarot pack which provide the theme, to rather incoherently spectacular effect. Even more than its predecessors this book exemplifies the interiorization of esoteric lore.

Technically the most flawless of this group of novels, *The Place of the Lion* (1931) describes what happens when a spiritual adept inadvertently opens the door between the worlds of matter and the archetypal energies which control it. The lion of strength, the serpent of subtlety, the butterfly of beauty, and their peers emerge into the Hertfordshire landscape and begin to draw the visible world back into the dimension in which it has its origin. Nowhere else does Williams make such effective use of supernaturalist motifs.

Down that provincial street all the horses of the world seemed pouring, but he realized that what he saw was only the reflection of the single Idea. One form, and only one, was galloping away from him: these other myriads were its symbols and exhalations. They were not there, not yet, how uneasily soever in stables and streets the horses of that neighbourhood stirred and stamped, and already kicked at gates and carts in order to break free. They were not yet there, although far away on Eastern and Western plains, the uneasy herds started, and threw up their heads and snuffed at the air, and whinnied, and broke into quick charges, feeling already upon the wind the message of that which they were. (ch. 10)

The centrality of spirit as a metaphysical category is here expressed in terms of body. The language, sometimes archaic, sometimes incantatory, is under the command of an intelligence entirely clear as to what it wishes to express.

Williams's most original achievements are his two final novels, *Descent into Hell* (1937) and *All Hallows' Eve* (1945). The former is an ambitious attempt to depict the natural world in its supernatural context. It portrays the mysterium at work not only in the providential ordering of human affairs, but also in the psychic and physical structuring of everyday life. The living and the dead cohabit space, only unaware of each other because of the passage of time; all the events in the book that really matter occur in the metaphysical dimension, as operations within the mysterium. A workman engaged in the building of a residential estate called Battle Hill commits suicide; a Protestant martyr is burnt on the same site; a girl encounters her *doppelgänger* and overcomes her fear of it. Events in the distant past, the recent past, and the immediate present happen simultaneously; but none of them, in whatever dimension they take place, is governed by the laws of clock time.

The book asserts its spiritual dimension in literal terms: there is no building-up of premonitory strangeness. The substance of Battle Hill is presented, almost naturalistically, as multi-layered, and the worlds of the living and the dead are shown to overlap and to be reciprocally influential and expressive. The encounter with the *doppelgänger* is narrated prosaically and given no pseudo-preternatural status. It only becomes momentous when its meaning is manifested, revealing its genuinely supernatural nature. Again, a demonic succubus appears—but not as an imposition or a visitation, rather as the deliberate creation of the victim's lust. Laurence Wentworth's quest for a masturbatory image to replace the girl who has preferred a younger man to himself is described as part of the interior universe in which all human thoughts and desires have their objective being. He hears the footsteps of the suicide, who in this spiritual limbo is searching for human contact and a place in which to be. Since Wentworth is himself a potential suicide, the workman serves as an image of his capability of repentance; but Wentworth rejects the idea of any intrusion upon his fantasy,

whose origin is with man's, kindred to him as he to his beasts; to whom a name was given in a myth, Lilith for a name and Eden for a myth, and she a stirring more certain than name or myth, who in one of her shapes went hurrying about the refuge of that Hill of skulls, and pattered and chattered on the Hill, hurrying, hurrying, for fear of time growing together, and squeezing her out, out of the interstices, of time where she lived, locust in the rock; time growing together into one, and squeezing her out, squeezing her down, out of the pressure of the universal present, down into depth, down into the opposite of that end, down into the ever and ever of the void. (ch. 5)

The incantatory, almost improvisatory nature of the breathless prose induces a sense of stifling self-enclosure, of self-destructive panic as the ancient myth is interiorized. One prominent feature of Williams's fiction is his confident handling of the imagery and mythological patterns of esoteric tradition, of the mystical theology that forms a common ground for the spiritual traditions of Judaism, Christianity, and Islam. He portrays the supernatural in terms not so much of psychic or physical phenomena as of spiritual law (in this respect benefiting from his earlier occultist studies). In his world of spiritual encounters, self-enclosure constitutes damnation: however unorthodox and unfamiliar in its iconographical expression, Williams's religious outlook is rigorous in its clarity of moral vision.

In *All Hallows' Eve* the overriding image is that of the city. Williams completed this novel shortly before his death. It has a testamentary quality, for in it he defines with forceful urgency his belief that what lies at the root of all magic is that 'the body was itself integral to spirit' (ch. 9). The book is a decisive refutation of dualism; it proclaims the sovereignty of the divine law and justice that causes all evil to play into its hands and thus refute itself. Even more than in *Descent into Hell*, Williams pitches his story within a spiritual dimension and shows the mysterium at work. A dead girl, through her developing self-knowledge and repentance, becomes the means of overthrowing the power of a charismatic magician, who is bent on making himself immortal —yet another embodiment of Simon Magus. *All Hallows' Eve* in its rejection of occultism is Williams's decisive vindication of a life of love and forgiveness over the hungry self-aggrandizement of magic and the quest for power.

The account of the girl's wanderings in the empty city of the dead is extraordinarily persuasive, perhaps because Williams fails to externalize the city, so that everything is mediated through her dawning self-awareness. The overlap of dimensions of reality, of which Williams is always so conscious, is captured in the description of a visionary painting and of its relation both to the world without and the world within.

He walked to the window and stood looking out. The grey October weather held nothing of the painting's glory, yet his eyes were so bedazzled with the glory that for a moment, however unillumined the houses were, their very mass was a kind of illumination. They were illustrious with being. The sun in the painting had not risen, but it had been on the point of rising, and the expectation that unrisen sun had aroused in him was so great that the actual sun, or some other and greater sun, seemed to be about to burst through the cloud that filled the natural sky. The world he could see from the window gaily mocked him with a promise of being an image of the painting, or of being the original of which the painting was but a painting. (ch. 7)

Williams was a true hermeticist. Few writers can so convincingly portray the immanence of divine energy and delight within the material order.

The range of his methodology and his belief in spiritual law, together with his refusal merely to sensationalize, make him one of the most persuasive of supernaturalist novelists. His work demonstrates the intellectual value of a firm dogmatic structure in any writing of this sort. Moreover, he was the only Christian imaginative writer of his time to have an informed understanding of the esoteric spiritual meanings that underlie the doctrines of formal theology, doctrines which can seem crudely literal when spelt out exclusively in chronological terms. Williams understood the fact of theological synchronicity, and accordingly subordinated the excitements of supernaturalist literary methodology to what may be called a metaphysical naturalism, as against the social naturalism that is a novel's customary concern. Indeed, of naturalism in the ordinary sense of external surfaces and social analysis he provides next to none, though the novels are steeped in a mental sophistication that offsets any tendency such a limitation may have towards the merely naïve. Thus although his portrayal of his characters' lives inclines to be offhand, and his diction highly mannered, these very idiosyncracies serve to display the momentous relevance of the themes that form the subject of his tales: in this case there is no question of a fictive naturalism embarrassing the claims to oracular significance. But the failure of these novels to attract a wide readership (they have maintained a small and enduring one) would suggest that the novel of magical, occultist, and preternatural properties remains marginal to the imaginative needs of twentieth-century readers. Even when used as intelligently as it is by Williams, the genre necessarily ignores the literary and philosophical implications of the erosion of belief in metaphysics.

In *A Passage to India* Forster makes a bitter reference to 'poor little talkative Christianity'.[8] In the context of the Marabar caves the comment seems self-evident: the evangelistic nature of this particular religion, and its presiding over the emergence of the hyper-rational, technologically manipulative civilization of the

West, has involved its ministers in an excess of explication and in retrograde attempts to accommodate demonstrable fact to prescriptive theory. Those writers who, during the critical hundred years after 1850, attempted to portray the workings of remedial providence against the background of evolutionism and the opening up of global communication (physical, mental, and cultural) were, in the very limitation of their achievements, to demonstrate the impermeability of the mysterium to adequate transcription.

Since theological novelists already possess a credal formulation as to the nature and function of the supernatural, their work necessarily excludes any note of agnostic inquiry or intellectual exploration, unless it be (as in the case of MacDonald and Williams) a matter of assuming a hermetic mantle in order to revivify the verbal expression of doctrines already dormant in the mind. MacDonald's use of supernaturalist material is more innovative in its imagery than it is linguistically; but his achievement in demonstrating the possibility of remythologizing the central doctrines of the Christian faith, is liberating both for the creed he served and for the literary methodology he used. Williams goes further still. He exploits a diversity of magical and religious iconography in order to universalize what was in as much danger of being constricted by its socially cohesive role as was ever any hapless young imaginative curate by the associations of 'the cloth'.

But if Williams was prepared to couch public proclamation in the tones of personal statement, it was in a mannered, arcane, and highly idiosyncratic idiom that necessarily limited his appeal. Certainly his failure to win any widespread popularity measures the difficulty, almost the impossibility, that a predetermined expression of belief should be able to convey the realities of the experience of mystery—one limitation that theological and hermetic writers have in common. Compared with the hermetic literary tradition, the theological one is expository rather than exploratory: it is in some surprising collocation or association of ideas that one gets the shock of supernatural surprise. Mainstream twentieth-century Christianity may have grown blind to metaphysics, but its supernaturalist writers continue to voice belief in an overriding spiritual order, and in doing so provide an exoteric interpretation of esoteric truth. It is perhaps their most important contribution to the literature of their time. Ironically, however, it remains true that they carry most conviction when they exhibit an awareness of how the secular world behaves and feels. An incarnational theology requires no less. In the case of MacDonald, Chesterton, and Williams, it is their spiritually enlightened worldly wisdom which authenticates their transcendental insights and assertions.

5

TWILIGHT TERRITORIES

> There's a noise far away in the dark which seems to know
> where I am.
>
> (Margiad Evans, *Autobiography*)

The turn of the century, which saw the beginning of the great age
of the English ghost story, was a time of simultaneous stability
and change. The stability was only apparent, the change was
real—the basic narrational premiss in fictions of this kind. Accordingly we find in the Edwardian era many a prominent novelist describing the economic, political, and sexual upheavals
which formed the undercurrent of that period—H. G. Wells,
Arnold Bennett, and E. M. Forster are obvious examples—while
the extent and the challenge of imperial power continued to be
celebrated, and questioned, by Rudyard Kipling (1865–1936), who
by the time Victoria died represented in popular regard the voice
of masculine enterprise and self-sufficiency. It was not at this
stage that the extraordinary sophistication of his artistry was to
be fully recognized; and most readers would probably have set
him alongside Wells, rather than alongside Henry James, in those
two writers' debate as to the supreme importance (or not) of an
artistic rendering of the novelist's material.

But if action, change, public affairs were one staple of Edwardian fiction, uncertainties, questioning, and private whimsy were
another: besides being the age of Kenneth Grahame and J. M.
Barrie the decade also saw the early work of Walter de la Mare
(1873–1956); and if one adapts James's metaphor of 'the house of
fiction' one might say that while all original and instinctive novelists inhabit houses of their own, the one indisputably haunted

one belongs to him. His reputation both as poet and as fiction-writer is based on his being associated with a universe of dream and fantasy and of childlike wonder (that faculty so beloved, and possibly invented, by adults disenchanted with the materialistic presuppositions from which they nevertheless remain insepar-able). On examination, the popular ascription does in part hold good: de la Mare's imaginative world is defended by thickets of ambiguity and, in his prose, of Jamesian stylistics. Kipling's house of fiction, on the other hand, appears at first sight only too ac-cessible, rackety as a barracks, the resort of men of action whose resourceful trickery is compromised by half-apologetic dewy sen-timent. But Kipling's house has its traps—dark corners, unex-pected mirrors in unflattering lights, mysteriously banging doors. His work, no less than de la Mare's, reflects the intense unease which a sense of a supernatural dimension engendered in the early twentieth-century consciousness as the confident values of Victorian materialism and religious fervour lurched towards the First World War. But this aspect of their work was to a large extent concealed by their popular reputations, and also by their readers' longing for tales of action and endeavour on the one hand, and on the other of glamour and escape. Both Kipling and de la Mare were ticketed by self-parodying readers who demanded of them caricatures of their peculiar gifts. The sharp divide be-tween the two categories with which they were respectively iden-tified is a mark of the decline in their age's understanding of the supernatural.

Rudyard Kipling

Under certain circumstances few hours are more entertaining than those spent in poring over caricatures of artistic and politi-cal celebrities, especially those devised by 'Max'. Instructive, too, for Beerbohm's portraits of (say) Henry James, while delineated with beneficent absurdity, actually enhance one's understanding of the style and method of *The Golden Bowl* and its immediate predecessors. The satire is not invariably benign, however, and artists of a quiet disposition seldom register such withering dis-like of an admired contemporary as does Max in his cartoons of Kipling. Others among his fellow writers did not get off lightly either—Oscar Wilde and Frank Harris come immediately to mind —but Kipling was pursued by Max with unrelenting detestation.

Cartoon after cartoon displays a beetle-browed wee man with goggle-spectacles and fierce moustache hobnobbing with John Bull across a pint of foaming beer or (still more deadly) taking 'a bloomin' day aht, on the blasted 'eath, along with Britannia, 'is gurl'. But although he hated what Kipling had done with his gifts, Beerbohm had to concede he was a genius: 'even *I* can't help *knowing* him to be that. The schoolboy, the bounder, and the brute—these three types have surely never found a more brilliant expression of themselves than in R.K.'[1]

Kipling's emphasis on masculinity Max found suspect, leading him to consider even the delicate and tender 'They' to be 'metallic. The house and garden and children all seem made of zinc. A ghost deserves better!' And indeed ghosts did well at Kipling's hands: his work in the medium of the supernaturalist story is outstandingly fine, even by comparison with that of his contemporaries. His attention to, and knowledge of, the kind of subject-matter not usually associated with writers of his calibre, and his concern with soldiers, politicians, industrialists, and business-men, scarcely prepare one for the subtlety and range of his supernaturalist tales. Yet in this department too he was a representative figure of his time. Many of his short stories deal with experiences on the psychic borderline, and tend to treat the supernatural as the paranormal. The Indian tales, understandably, incline more to the preternatural, an alien culture begetting a sense of the eerie and the strange. The later stories set in England, however, shade into a feeling for the supernatural in the fullest sense.

The Indian tales focus on the lives of the expatriate English, and the ghost stories, so to call them, reflect the strains imposed by climate and the arduousness of administration. 'The Mark of the Beast',[2] for example, a werewolf story with an uneasily sadistic tone, portrays the consequences when a drunken Englishman defiles a Hindu temple. The hideous conclusion to these bizarre events is not accounted for, nor does the narrator expect it can be, 'because, in the first place, no one will believe a rather unpleasant story, and, in the second, it is well known to every right-minded man that the gods of the heathen are stone and brass, and any attempt to deal with them otherwise is justly condemned'. In view of what has gone before, the irony resounds well beyond its immediate occasion.

Still more impressive is 'At the End of the Passage',[3] which has some claim to be regarded as Kipling's 'Heart of Darkness'. The

stresses endured by a quartet of Englishmen who forgather each
week during the hot season in the plains, makes it possible for
him to focus more on the fear aroused in the victim of the preternatural visitation than on the visitation itself: the reader is
infected through contagion rather than through any uncanny
spectacle. The horrors of self-induced sleeplessness are even more
real than the horror which may assault the sufferer should he not
stay wakeful—a 'blind face that cries and can't wipe its eyes, a
blind face that chases him down corridors'. If the photographing
of the dead man's eyes to find out what he has seen is an instance
of Kipling's attraction toward the paranormal, the ending, when
the photograph is developed and the result immediately destroyed,
anticipates the tales of M. R. James. But the story is rooted in the
Englishman's physical experience of India. In so graphically evoking the agonies of helpless terror and the sense of immersion in
a foreign and mysterious folk culture, 'At the End of the Passage'
is unusual among tales of the supernatural, not for its horror but
for its depth of pity.

Other stories are more conventional, like the better known 'The
Phantom Rickshaw',[4] which lies open to interpretation as the tale
of a delusion; and it is in fact less creepy than the account of a
real delusion, 'My Own True Ghost Story',[5] in which, as in 'At the
End of the Passage' the effect depends on the experience of fear
itself. These tales shade into others which describe the weird but
not the inexplicable, tales such as 'The Strange Ride of Morrowbie
Jukes',[6] with its hideous picture of those living ghosts, the ritually dead who are still alive; or 'The Return of Imray',[7] an account of psychic disturbances and the extrasensory perceptions
of a dog. They form an integral part of the writer's responses to
the lives of 'mine own people'.

The same thing might be said concerning the supernaturalist
element in some of Kipling's stories of the First World War. In
'A Madonna of the Trenches' and 'The Gardener' the experience
of bereavement is deepened to a metaphysical condition. The
former story,[8] which describes the apparition of a dead woman
to her lover in the trenches, is told with all the indirection of
which Kipling was a master: the actual encounter appears as a
tale within a tale within a tale. The fact that one is shown the
meeting through the appalled eyes of the woman's nephew, a
Cockney runner who is used to treading on the frozen faces of
the 'stiffs' employed as protection against the mud, makes its

own comment: a boy who can stomach the debasement and obscenity of physical death is distraught at the sight of a love which can transcend it. The tale is at once the most original of ghost stories and a bitter comment on a world which so easily countenances the conditions human stupidity imposes to its own undoing. The title, though provocative, is not ironic.

In 'The Gardener'[9] a woman who has passed off her illegitimate child as her nephew, keeping the truth even from him, visits his grave in Flanders, still without giving away her secret: a young gardener shows her the place where he is laid, and refers to him as her son. The ascription could be a human error; but the final sentence of the story, with its quotation from the Resurrection narrative in St John's Gospel, permits no doubt as to the identity of the speaker, even though the woman herself leaves with her eyes unopened. Like so many of Kipling's tales it is one which a touch less deft would have rendered crass; but even on successive readings the effect is of an authentic intuition as to the mysterium.

Other stories deal with the paranormal. 'The Finest Story in the World'[10] combines an account of a young man's unconscious dreaming, which takes him into the past, with a humorous and satirical picture of the inability of unimaginative people to perceive the riches around them: it relates pleasantly to Kipling's own stories in the Puck books. In 'Wireless'[11] a chemist's consumptive young assistant is found in a trance composing 'The Eve of St Agnes', while in the back part of the shop an amateur radiographer is picking up signals on the ether. The thesis of the tale does not bear examination, but the enthusiasm with which the author recounts his story is irresistible. At the back of all his supernaturalist tales of whatever kind is his knowledge of the universe's unfathomable and many-sided character.

It is a celebration of provincial life, however, which marks Kipling's finest achievements in this genre. A love for Sussex suffuses the Puck stories with all the naturalness of one who is at home with his readers. Something of his feelings for Batemans, his seventeenth-century house near Burwash, can also be gauged from two other stories published at this time. 'An Habitation Enforced'[12] tells of an American couple's gradual adoption by the countryside they think has been adopted by themselves; the story's seemingly artless conservatism, while unquestioned, is indispensable to the irony of the whole. 'The House Surgeon'[13] is a tale of the paranormal, of a haunting rooted not in the malignity of the past but in the unhappiness of the present, as a woman's suspicion that her sister may have killed herself holds back an innocent spirit that longs only to be laid to rest. In both tales the supernatural acts upon human beings and is part of them: Kipling plays the part of a spiritual anthropologist. But it is in 'They', 'The Wish House', and the Puck stories that his powers as a rural supernaturalist are most effective.

By virtue of their being books for children, the latter are usually placed in the category of fantasy; but the fantasy on which they draw amounts to myth. In the opening chapters of *Puck of Pook's Hill* (1906) Kipling has his children perform a miniature version of *A Midsummer Night's Dream* in a fairy ring at a bend in the mill stream.

Three Cows had been milked and were grazing steadily with a tearing noise that one could hear all down the meadow; and the noise of the Mill at work sounded like bare feet running on hard ground. A cuckoo sat on a gate-post singing his broken June tune, 'cuckoo-cuk,' while a busy kingfisher crossed from the mill-stream to the brook which ran on the other side of the meadow. Everything else was a sort of thick, sleepy stillness smelling of meadow-sweet and dry grass.

It is the quintessential turn-of-the-century England, the England that the coming of the motor car was going to transform into an object of nostalgia, the England conjured up in various ways by Hardy, Richard Jefferies, Kenneth Grahame, and Edward Thomas and evoked in plangent music by so many early twentieth-century composers. Puck himself comes from the same imaginative world as Edward Thomas's poem 'Lob'. But Kipling gives his scene a specific historical location. In *Something of Myself* (1937) he tells of his first impressions of the country round Batemans, with

the long, overgrown slag-heap of a most ancient forge, supposed to have been worked by the Phoenicians and Romans and, since then, uninterruptedly till the middle of the eighteenth century. The bracken and rush-patches still hid stray pigs of iron, and if one scratched a few inches through the rabbit-shaven turf, one came on the narrow mule-tracks of peacock-hued furnace-slag laid down in Elizabeth's day. The ghost of a road climbed out of this dead arena, and crossed our fields, where it was known as 'The Gunway', and popularly connected with Armada times. Every foot of that little corner was alive with ghosts and shadows.[14]

It was out of this recognition of a living past that the Puck stories were created. Puck is the last of the Old People, the so-called fairies, and his earthiness proclaims him for what he is, an earth spirit and, for Kipling, specifically that of the English earth. His natural magic enables the children to meet people who have lived and worked the landscape round them. They are not, as are E. Nesbit's children in *The Story of the Amulet* (published in the same year as *Puck*) transported into the past: the past comes to them. The magic of oak, ash, and thorn protects them from the demands of their own world: Kipling here supernaturalizes the life of his native country without intellectualizing it or evoking— to use Coleridge's term—a timeless 'Platonic England'.[15] The fact that a number of the stories (most notably 'Dymchurch Flit'[16] and 'Cold Iron')[17] contain preternatural elements within their already supernaturalist frame only serves to emphasize Kipling's integration of such beliefs into his presentation of human life as a whole.

His treatment of the children is entirely natural: Dan and Una are clearly his own John and Elsie. But it was a third child, Josephine, who died in 1899, who was the inspiration of 'They'.[18] The unnamed narrator, proudly driving his new motor in the Sussex lanes, comes across an old house and garden inhabited by a blind woman and, apparently, by numerous children. As the story unfolds it becomes evident that these are dead children who cannot bear to leave the earth, and whom the childless woman's love has brought back for safe keeping: one of them, as is made evident in a moment of great poignancy, is the narrator's own.

It is a story which could easily have been sentimental. Kipling was, after all, writing at the same time as J. M. Barrie and Richard Middleton, as Kenneth Grahame and Forrest Reid, all of whom were presenting children with a charm that bordered upon whimsy: Middleton's story 'The Passing of Edward' has distinct affinities with 'They'.[19] But Kipling's tale is distinguished by an accompanying robustness, evident in the narrator's enthusiasm for his motor car; in his occasional facetiousness (which helps to point up the other-worldly elements, rather as Lockwood's crassness enhances the effect of the hauntings in *Wuthering Heights*); and in the unflattering portrayal of the village people, even in the sense of 'the county'. The house is well known to the community around it, and a newly bereaved mother is said to 'walk in the wood': a residual paganism is suggested here. And the blind woman herself is tough: perhaps the most telling moment of all is when the narrator makes contact with his child while his hostess is hard-headedly negotiating with a greedy tenant, for whom the presence of the unseen children is an experience of preternatural dread. The narrator's departure at the end of the story is appropriate and right; he must not cling to what has to be taken from him. The child may fear the separation, but he himself must not. The nostalgia of Kipling's contemporaries is rejected. The house is for the children not for men. The story carries in it the very essence of bereavement, softened only slightly, but controlled.

This sense of ongoing human suffering introduces a similar vein of compassion into 'The Wish House'.[20] Two elderly country women sit reminiscing in a cottage parlour, while the traffic thunders by and a football match is being played; one of them has cancer and the other is going blind, and their meeting is broken up by a routine visit from the district nurse. It is the sort of vignette found in the contemporary work of Katherine Mansfield or Malachi Whittaker, a little broader in treatment (the old ladies get some of their words wrong) but adroit in the way the author reveals their past histories through allusion and the occasional clash of personalities. Into these exchanges there comes the account of the visit of one of them to a 'Wish House', 'a house which 'ad stood unlet an' empty long enough for Some One, like, to come an' in'abit there'. This someone is 'a Token . . . a wraith of the dead or, worse still, of the living'. The Wish House is, by a master-stroke, a terrace house in an ordinary London street; and there the woman delivers her wish that she may take on 'everythin' bad that's in store for my man'. The wish is granted; her later one, that he at least will marry nobody else is granted too, though for a more prosaic reason; he has a possessive mother, 'an' as long as she was atop the mowlds, she'd contrive for 'im till 'er 'ands dropped off. So I knowed she'd do watch-dog for me, 'thout askin' for bones.' The humour only serves to reinforce the story's credibility; what matters is the power of the woman's belief and her ability to live by it. But the brief moment of pre-ternatural terror is neatly managed. What the woman hears as she delivers her message is 'feet on de kitchen-stairs, like it might ha' been a heavy woman in slippers'—in other words an image of her present self. A little girl, on the other hand, only hears gig-gling when she goes to the Wish House door.

Like another of Kipling's rural tales, 'Friendly Brook',[21] 'The

Wish House' is as much about belief in the supernatural as about the supernatural itself: the whole idea of the Wish House is a matter of folklore—the little girl regards it as a kind of sacred game. But the strangeness of the belief, and of the episode as recounted, suffuses the whole story, dignifying the exchanges between the old women, lending an air of mystery to a tale which the author is at pains to keep as naturalistic and integrated with contemporary life as possible. As with 'A Madonna of the Trenches', the preternatural element is allied to the power of selfless human love.

Kipling's contribution to the literature of the supernatural is marked by its comprehensiveness. The preternatural elements are variously depicted as malign, wayward, aboriginal, spiritual, specifically Christian, and redemptive. They reflect a shared experience of popular belief rather than the communings of the lonely self. Because he does not explore the metaphysical and psychological questions which an experience of the supernatural can arouse, Kipling, despite the brilliance of his individual achievements in this vein, is not in the true sense of the term a supernaturalist writer.

Walter de la Mare

That ascription applies most certainly, however, to Walter de la Mare, a target for Max's more benignant mockery. The middle-aged author, chin propped on hands as he crouches on a footstool, gazes up at an elderly seated woman (in bootees such as only Max would draw) 'gaining inspiration for an eerie and lovely story'. As with all Beerbohm's literary cartoons, as well as being a witty piece of portraiture the picture is subtly critical of its subject; but unlike the various sketches of Kipling, it is not unfair. Drawn in 1925, it reflects de la Mare's literary reputation at that time and as it has to some extent remained. The cartoon mocks the readership rather than the writings.

By then de la Mare had published all four of his full-length prose fictions,[22] as well as two collections of short stories; but he was still popularly known as a writer of verse. Merely fanciful and charming though some of it is, his poetry displays a preoccupation with metaphysical experience that stamps him as being the last voice (some might say the last gasp) of the Romantic tradition. The poems are full of disquiet, and express a defeated longing for some revelation of mystery behind the visible universe, for some reinforcement of a belief in the supernatural that was currently being eroded. Their conclusions are bleak: in answer to a cry for reassurance,

> ... the silence surged softly backward,
> When the plunging hooves were gone.[23]

And with still greater desolation,

> Nought but vast sorrow was there—
> The sweet cheat gone.[24]

'Gone', 'gone' ... This bitter frustration is explored in de la Mare's prose writings also, and with a thoroughness, variety, and complexity which the author's popular reputation regrettably obscures. As a poet de la Mare is an enduring but minor voice; as a writer of ghostly tales he is among the masters of the craft, unmatched in his ability to evoke the twilit territory of psychic borderlands.

Characteristic of this gift is the story 'Winter'.[25] In it the narrator, deciphering at dusk the headstones in a snowbound country churchyard, encounters a being who regards him 'with unconcealed horror'. Moreover, 'this being, in human likeness, was not of my kind, nor of my reality'. Its voice comes 'as if from within rather than from without' the narrator's ear, and its beautiful face has 'almost colourless eyes and honey-coloured skin'. The delicate precision of the writing only serves to reinforce the strangeness it purveys. The story is an evocation of preternatural experience that is entirely self-sufficient.

A sense of greater moral urgency, however, informs de la Mare's most elaborate study of the supernatural, *The Return* (1910). It too opens in a country churchyard. Arthur Lawford, an indolent middle-aged husband and father, drops asleep beside the grave of an eighteenth-century suicide called Nicholas Sabbathier—to find on his return home that the latter has taken possession of his physical appearance, and is in process of taking over his mind and soul. It is characteristic of de la Mare that the body should succumb more readily than the soul: it is as though a parable of a psychological condition were being written in terms of a tale of the preternatural—a reversal of traditional literary procedure. But *The Return* is more complex than that. Lawford's ultimately victorious struggle with the spiritual invader is internalized; at the same time Sabbathier's presence is made known not only to him but to others, notably a housemaid who picks up its preternatural vibrations. Moreover, his condition is recognized and taken seriously by a bachelor recluse, whose sister Grisel is an incarnation of Lawford's 'ideal beloved'. She and Sabbathier have been lovers in the realm of the spirit from which she sends each of the two back to his proper place—which in Lawford's case is the world of his unimaginative wife Sheila, who deeply resents the unaccustomed experience into which she has been thrust. This return is to a state of exile, the materialistic world in which men and women are stultified and incomplete: it is necessary but painful, and the novel registers the pain rather than the need for pain.

The return is one in which two worlds intermingle. The book is thick with ambiguities, furthered by an elaborate, meticulously orchestrated prose.

The flaming rose that had swiftly surged from the west into the zenith, dyeing all the churchyard grass a wild and vivid green and the stooping stones above it a pure faint purple, waned softly back like a falling fountain into its basin. In a few minutes, only a bar of orange burned in the west, dimly illuminating with its band of light the huddled figure on his low wooden seat, his right hand still pressed against a dully beating heart. Dusk gathered; the first white stars appeared; out of the shadowy fields a nightjar purred. But there was only the silence of the falling dew among the graves. Down here, under the ink-black cypresses, the blades of the grass were stooping with cold drops; and darkness lay like the hem of an enormous cloak, whose jewels above the breast of its wearer might be in the unfathomable clearness the glittering constellations . . . (ch. 1, author's ellipsis)

The influence of Meredith is obvious; but one notes also the transference of the epithet 'stooping' from stones to grass, the disturbing verbal incongruity whereby a bird is said to purr, and also the movement from precise, directly presented observation to a simile that recoils upon itself, fusing the two worlds of sight and imagination. It is in this twilight world, always magical, that the spirit of Sabbathier manifests itself and, by an irony which is the book's main theme, restores to Lawford his capacity to appreciate the life to which the suicide has returned, and of which he is to try to take possession. For Lawford the process is not one of loss, for 'he was conscious of having learned in these last few days: he knew what kind of place he was alone *in*'. 'It was this mystery, bereft now of all fear, and this beauty together, that made life the endless, changing and yet changeless, thing it was. And yet mystery and loveliness alike were only really appreciable with one's legs, as it were, dangling down over into the grave' (ch. 23).

Here is de la Mare's definition of mystery: the grave and all it signifies holds a permanent fascination for him. But *The Return* envisages a more positive outlook than this preoccupation might suggest: it moves from awareness of the preternatural to acceptance of the visionary. The most intense moment in the book comes when Lawford sees with Sabbathier's eyes the inner landscape which he shares with Grisel. Its source is a farmhouse, a pool, a village. Gazing at them,

it seemed as if a thin and dark cloud began to be quietly withdrawn from over his eyes . . . he was staring in the stagnancy of a trance at an open window against which the sun was beating in a bristling torrent of gold, while out of the garden beyond came the voice of some evening bird singing with such an unspeakable ecstasy of grief it seemed it must be perched upon the confines of another world. The light gathered to a radiance almost intolerable, driving back with its raining beams some memory, forlorn, remorseless, remote. His body stood dark and senseless, rocking in the air on the hillside as if bereft of its spirit. Then his hands were drawn over his eyes. He turned unsteadily and made his way, as through a thick, drizzling haze, slowly back. (ch. 20)

Grisel recognizes the experience for what it is. She answers his longing with 'Don't think; don't even try to. It mustn't be. We can't; we *mustn't* go back.' The words are applicable to all attempts to break through the restrictions of time and space—if restrictions they be—and they voice the essential difference between true supernaturalism and the occult.

Surprisingly enough, only eleven of de la Mare's short stories feature a ghost in the strict sense of the term; and although excellent examples of their kind, these have by their very nature a certain definitiveness alien to their author's inspiration. In 'A Revenant'[26] the spectre of Edgar Allen Poe attends a lecture on his life and work; but the real burden of the story is a satire on the plundering of the lives of the dead by speculative biographers. 'The Green Room',[27] which concerns the discovery and publication of a cache of passionate poems written by the suicide who haunts an old bookshop, has a not dissimilar point. Neither story is designed to frighten in the manner of M. R. James. The most creepy of de la Mare's tales in this kind is 'Bad Company',[28] in which the ghost of an evil old man encounters the narrator on the London Underground, and lures him back to the house where his corpse is sitting by itself—a moment of authentic horror. It is the reaching out of the dead to the living across the invisible barrier that is de la Mare's recurrent preoccupation.

For it is less the haunters than the haunted who interest him. His world is crowded with furtive, guilty people, as unsettling as they are piteous, among them being the seedy stranger in the station waiting room in 'Crewe',[29] the lonely occultist in 'A Recluse',[30] and Seaton's abominable aunt.[31] And this world is full of physically frustrated, stunted beings, like the young man in 'At First Sight',[32] unable to raise his eyes above the ground, and robbed of his only chance of love by an overly protective grandmother. Ruthless elderly women proliferate: the bleakly self-deluded and imperceptive aunt who narrates 'The Guardian',[33] the more sympathetic mother of the mad failed poet in 'Willows',[34] or the repressive Miss Jemima.[35] This last story is especially interesting for its double focus. An old lady recalls to her small granddaughter the persecution she endured as a litle girl from her uncle's housekeeper, and her own entrapment by a fairy in a churchyard. The fairy is an elemental: and there is an eerie moment when the child, hidden from Miss Jemima in the church, hears her footsteps pause on the threshold and realizes that she too is aware of the fairy and is afraid to venture further. Evil calls to evil—or, say, unnaturalness to unnaturalness; nor is the child in her hatred guiltless either. The story, professedly for children, generates a characteristic disquiet.

At times the old women and the children change places. In 'Miss Duveen',[36] a crazed creature confronted by the small boy narrator is viewed by him with a heartlessness not so far from that of her sadistic guardians. Indeed, de la Mare is singularly unsentimental as to the callousness of children: 'In the Forest',[37] 'An Ideal Craftsman',[38] 'The Bowl'[39] come to mind, though the last-named story gains its effects partly through its realization of the knife-edge trodden by a child between an extreme of sensibility and the carapace of limited perception. But in de la Mare's children and guardians alike there is often a frightening blindness to the presence of mystery: they sense it, but they will not look. Much of his most poignant writing is concerned with this jerking away, itself a reaction characteristic of the contemporary materialistic outlook that determines so many of his people's lives.

Those who do see, who will gaze, are confronted with beauty, desolation, and terror. In 'All Hallows'[40] an old verger conducts the narrator round a lonely decaying cathedral near the sea, hinting that it is being rebuilt by demonic forces—and then what is whispered about and felt is seen and heard. The child in 'Physic'[41] wants the blinds drawn down: 'why do faces come in the window, horrid faces?'. The mother in 'The Wharf'[42] dreams of the world as a refuse dump where dead souls are shovelled away indifferently by God's angels; but this story is resolved by the vision of a healthy farmyard dung-heap, one of de la Mare's very few unambiguous imaginative devices. But such a transformation is implicit in the alternative vision put forward in 'The Creatures' and 'The Looking Glass', which are among the most sure and subtle of their author's achievements.

In 'The Looking Glass'[43] a consumptive young girl is told in an offhand way by her employer that the garden of her house is haunted. The story is taken up by the cook next door, a gloomy woman, completely materialistic in her approach to psychic phenomena. She encourages Alice's superstition, but offers nothing beyond it: 'it's all death the other side—all death'. Alice perceives that 'the other side' may be a delusion: the garden is like a looking glass, which conceals all that is behind and beyond it, returning only 'the looker's wonder or simply her vanity, even her gaiety'. Solipsism is robbed of its sting: Alice's final realization that 'The Spirit is *me*: *I* haunt this place' is a source of joy. Her death on the very May Day Eve on which she has planned to make a ritual-magic visit to the garden, in order to encounter the mystery for herself, is accordingly a logical foreclosure.

'The Creatures'[44] transmutes the traditional supernaturalist tale still more radically. As so often with de la Mare, the story is told to the narrator by an enigmatic companion. The stranger, encountered on a train, tells of a chance visit paid to an earthly paradise while he is out on his 'goal-less wanderings' in lonely country near the sea (the same kind of country described in the more sinister stories 'Mr Kempe'[45] and 'All Hallows'). On a farm in a fold of the hills he finds a recluse and two strange beings whose happy simplicity transfigures their unsightly dwarfish appearance and 'shrill guttural cries' into simulacra of Eve and Adam in this wild Eden full of unfamiliar vegetation and idyllic silence. Then, later, in a nearby inn, the nameless stranger is informed that the creatures are retarded, offspring of the farmer and a woman 'from over the sea'. The deconstruction of the vision is followed by a visit to the woman's grave. The surname of the farmer is 'Creature', the woman's name is 'Femina'. The mythical is given back, the vision restored, but still in terms of the tangible and accountably real. Yet the final effect is disturbing, consolatory but ambiguous. The stranger's story (which concludes the tale, for we are not returned to the original narrator) is thus genuinely mysterious.

De la Mare's understanding of mystery is one of limitless uncertainty. His mind is speculative, not logical; the long meandering disquisitions on time, memory, dreams, the imagination, which fill the introductions to his anthologies, are to some mentalities maddeningly unstructured, to others totally beguiling. At its worst his style can be whimsical, evasive, arch—as Graham Greene points out, this particularly tends to happen when he assumes a feminine persona[46]—but again and again one is brought up short by some adroit intensity of phrase, by some flash of insight or arresting simile. He is never dogmatic, occultist, or the maker of any precise distinctions between the metaphysical and the material worlds. If his own world is made of moonshine, it is a real moon which shines, and the world it illuminates is not fantastical or figurative but a reality which is tangible and can be known.

As a result, his literary handling of the mysterium is extraordinarily supple. It is most usefully described in relation to what he designates the co-conscious. He provides no clear-cut alignments or different orders of reality, no abrupt confrontations between normal and abnormal, natural or preternatural. Even in the formal ghost stories the apparitions are either mistaken for people in the material world, as they are in 'A Revenant' or 'Strangers and Pilgrims';[47] or, as in 'Out of the Deep',[48] are manifestations of the beholder's psyche; they do not stalk in clanking chains like the ghost of Marley in A Christmas Carol or even with the soft padding thump of Le Fanu's invasive creatures. De la Mare, who frequently resorts to Coleridge's term 'my shaping spirit of Imagination'[49], constructs the spirits in his stories out of the imaginative lives of those encountering them.

Some of his most memorable portraits are of people trapped on a borderline raised in their minds by the separation of spirit from matter, those who are in quest of a materialization of the spiritual. Such are the crazed Mr Kempe, obsessed to the point of sadism with capturing the soul at the moment of departure out of life; or the wretched Mr Bloom in his deceitfully vacant-looking house called Montresor. 'This house was not haunted, it was infested.' Indeed, 'A Recluse' is as terrifying a story, even at the spectral level, as anything imagined by M. R. James. Even so, it is pathos rather than terror which constitutes the final note as the brash young narrator recalls the 'dark, affectionate, saddened, hungry eyes' of his tormented host.

De la Mare shows marked compassion for such heretics of the spirit. Their obsessions are offset by the quest for a spiritual other-world of which the guardian, pledge, and summoner is the idealized female figure, such as Grisel in *The Return*, who appears in a number of the tales. She is the messenger from beyond, the voicer of the truth—most memorably in 'The Bird of Travel',[50] where the dying Elizabeth proffers what is de la Mare's most stable underlying consolation that 'What is space but the all I am? What is time but the all I was and shall be?' To call such solipsism a conviction is perhaps too strong; but it is one on which the unceasing questioner comes either joyfully or wearily to rest.

In 'The Face'[51] and 'Lucy',[52] however, a female countenance seen in a pool becomes the revelation of beauty, love, and joy, which, far from being unattainable, is the beholder's own. De la Mare presents love as a transcendental experience, a gateway to Paradise (though whether Paradise lost or Paradise gained is not always clear); and his idealized female figures have as little vitality as those in a painting by Burne-Jones. Most of his men, on the other hand, are frightened, perverse, obtuse, or impotent. He shares the Edwardian preoccupation, so evident in the work of E. M. Forster and L. P. Hartley, with the sickly, limited, inhibited young man, dominated by mother figures and powerless to obtain his female counterpart, the anima. Several of the stories reveal a hopelessness that verges on despair. In 'The Picnic'[53] a lonely shop assistant falls in love with the face of a man seen through an upstairs window in the resort where she is spending her solitary holiday. Planning to meet him, convinced that he is a soul mate as alert to her as she to him, she finally encounters him—to realize that he is blind.

Joy and anguish are balanced in de la Mare's finest supernaturalist tales, as they are in those of Kipling. Both writers concern themselves most particularly with the point of contact between the external and the metaphysical worlds: theirs is an existential exploration of the psychic borderline, a twilight territory through which it is dangerous to venture into the realm of the supernatural. There is no magic circle of doctrinally defined religious faith to protect them. Kipling's natural curiosity gives his work a saving detachment that in no way diminishes its authenticity: his inclination is towards a reading of preternatural experience as being paranormal. The world he moves in has its demarcations and the sensitized consciousness retains its firm identity. But de la Mare's enquirers are engulfed in a world they cannot control. Dylan Thomas describes it well in suggesting that its ghosts 'though they reek and scamper, and in old houses at the proper bad hours are heard sometimes at their infectious business, are not for you to see. But there is no assurance they do not see *you*'.[54] De la Mare's imaginative vision is made up of hints and shadows in a retarded late nineteenth-century half-light, often stifling and confusing, yet taxing to both intellect and spirit in its very enervation. A face seen in a pool is the closest revelation of transcendence that it has to offer.

6

NUMINOUS LANDSCAPES

An angel satyr walks these hills.

(Francis Kilvert, *Diary*)

When Sir Walter Scott dispatched young Edward Waverley to
the Highlands he effected a wholesale conversion in the expectations of English novel-readers. The popular success of the
romantic narrative poem *The Lady of the Lake* (1810) had
familiarized his public with the charms of Loch Katrine; but
Waverley (1813) and its successors were to impart the aura of
romance, and an identification of mood and emotion, to a wide
variety of other regions, regions to be regarded no longer just as
scenery, but as climate, geology, and history to be experienced
and instructively enjoyed. Thereafter, a sense of place became a
significant element in the composition of prose fiction.

The posthumous publication in 1850 of Wordsworth's *The
Prelude* was, from this point of view, well timed, for the poem
explores the reciprocal relationship between a landscape and its
beholder. Around the middle of the nineteenth century the choice
of appropriate fictional settings and environments becomes increasingly important to a number of novelists: Anthony Trollope,
Elizabeth Gaskell, and George Eliot all possess a developed sense
of topography, one shared by Dickens, who in *Dombey and Son*,
Bleak House, and *Great Expectations* portrays particular localities
with a haunting sense of their symbolic and formative qualities;
while the use of sombre landscape by Wilkie Collins develops the
scene-painting of Ann Radcliffe into an oblique comment on the
destinies of individual characters and of their fortunes in society.
But it is in the novels of Hardy that the concern with place and

with particular regional characteristics comes to full fruition; he especially emphasizes the importance in his people's lives of rootedness, and the tragic waste involved in the increasing dislocation of society resulting from industrialization. With this elegaic aspect he stresses inevitably the beauty and well-nigh numinous significance of what was being lost.

But that numinosity was not merely a matter of nostalgia. Hardy shared in the nineteenth-century discovery of the pictorial values of landscape, evident in the paintings of Constable, Turner, and their many imitators, as a matter not merely of composition but of surface textures and of light: and the master-interpreter here was Ruskin, whose *Modern Painters* (1843–60) is in more senses than one an aesthete's bible. To its author, nature itself was a sacred book to be deciphered and analysed and, where possible, imitated. A later, more materialistic, age may interpret such an attitude to art as sentimental ideology; but it is less easy to realize that what it amounted to was a reading of the visible world as a vehicle for values of the spirit, and as being both the exhibition and the demonstration of a realm of supernature that was not at odds with the material universe, but which dwelt within it and fulfilled it. A landscape was no longer a mere backdrop to a group of figures; rather, it shaped them and shared in that of which they were made.[1] Surfaces were no longer a veil for the unseen; they were alight with potential sanctity.

There was, however, a limitation to this belief, one of which Ruskin himself was acutely conscious. Those surfaces were not those of the man-made industrial order nor of the majority of the artefacts it spawned. To this extent, nature remained opposed to man, as alien aesthetically as he in his turn was alien from nature through the possession of a moral sense. The network of duties and obligations which made up the ethical pattern of nineteenth-century Protestantism was counterbalanced by the concept of a journey from this dark and fallen world to the heavenly country awaiting the pilgrim beyond death. The world of nature was there to indicate waymarks and exemplary models: from it one primarily *learnt*. For Ruskin, beauty inhered in what was tangible and knowable; accordingly he supported the naturalistic ideal, maintaining that 'All great art represents something that it sees or believes in;—nothing unseen or uncredited.'[2] To appreciate landscape accurately was to behold and study its creator's hand in it.

For space, as much as time, gives entry to the mysterium; and the inherent values implicit in the slower pace and relative non-competitiveness of country life, as against the perpetual mobility and fragmentation of industrialized existence, acquired something approaching a religious significance for those who embraced what was felt to be the natural and divinely appointed way for humanity to live: Cowper's assertion that 'God made the country and man made the town'[3] became the most acceptable of clichés. By the early twentieth century the Arts and Crafts movement, and writings such as those of Edward Carpenter, were furthering a sense of the inherent healing qualities of natural scenery, a scenery made increasingly accessible by the railways and, in due course, by motor vehicles—modes of transport whose manufacture involved the very conditions that the new ruralists deplored. The paradox did not prevent a growing sensitivity to the imaginative suggestiveness of specific landscapes; while the possibility of weekend freedom, because of its very brevity, made those landscapes more romantic, more poignant, and more mysterious. Despite the efforts of some rural novelists to prove otherwise, this kind of sensibility had little to do with the actual work involved in cultivating the land; but to the townsman, whether visiting or retired, the Sussex of Kipling's imagination, together with other English counties (Housman's Shropshire most notably), being local and indigenous, unlike the foreign sublimities of Ann Radcliffe, began to function as metaphors for a homeland of the spirit. Fictive landscapes acquired numinous qualities which awakened in those who responded to them both inward and outward dimensions of an experience which could be termed reverential, if not religious. Wordsworth's intimations as to the reciprocity between the energies of nature and the capacities of the human imagination reappear in the shape of a renewed sense of the supernatural as being apprehensible within the visible world and not merely external to it. But that presence was often portrayed in ambiguous terms. Even among official professors of religion the underlying assumptions of rationalistic Deism were prevalent enough to account for that.

The Elusiveness of Pan

A present-day motorist who, for legendary association's sake, glances to the left on taking the Kendal road from Skipton, will see the colossal hulk of Pendle Hill rearing over the moors that form the distant boundaries of Lancashire. Pendle is the kind of isolated height that rivets the attention and gives birth to legends: so assertive a presence must, it would seem, have some occult or mythical significance. And the hill does have both its historical and its legendary associations. Since the seventeenth century it has been popularly connected with the outbreak of diabolism that forms the subject of what is probably the earliest regional supernaturalist novel, Harrison Ainsworth's *The Lancashire Witches* (1848). But it is the regional rather than the supernaturalist aspect of that book which lends it credibility today.

Ainsworth was a Manchester man, who knew the Forest of Pendle intimately and responded to its particular terrain. He had already employed the Gothic tradition successfully in *Rookwood* (1834), and was familiar with the tricks of the trade; but the result in the later novel is a kind of literary schizophrenia between an awareness of the indigenous nature of the supernatural as experienced by particular people in the places where they live, and a second-hand bookish insincerity, dictated by an aquiescence in literary fashion. Accordingly, whereas the bearded, four-square, physically hardy old witch, Mother Demdike, is a convincing product of her environment, her rival, Mother Chattox, remains the decrepit cackling crone of literary cliché. But in their close involvement with the lives of the local people, both women are more convincing than are the various demons and familiars which sound-off, appear, and disappear, like creatures in a pantomime. The author clearly believes in none of them; and the fact that these properties can be confounded by the mere mention of the word 'Salvation', let alone the pious euphemism 'Heaven', betrays the early nineteenth-century Evangelical provenance of his outlook. There is a theological discrepancy between the book's symbolism and the abstract vocabulary purveying it. What does linger in the mind is the character of the malevolent child, Jennet, a rare case in Victorian times of the presentation of juvenile depravity; and the accounts of Pendle Hill and the surrounding country, of the hunting of deer and otter, and an overall northernness, whose quality calls out for interpretation in more than simply naturalistic terms. *The Lancashire Witches* is no *Wuthering Heights*: its physical and numinous constituents are compartmentalized. Even so, however disappointingly, the novel does anticipate the presentation of idealized and significant fictive landscapes by writers with a more searching interest in discovering how a novel about rural life could accommodate a valid concept of the supernatural.

Thomas Hardy, however, is not among them, the romantic aspects of his novels being self-confessedly the product of his emotions. As he remarks in the Preface to *A Pair of Blue Eyes* (1873), the landscape round 'Castle Boterel' (Boscastle, on the north Cornish coast) is 'the region of dream and mystery. The ghostly birds, the pall-like sea, the frothy wind, the eternal soliloquy of the waters, the bloom of dark purple cast, that seems to exhale from the shoreward precipices, in themselves lend to the scene an atmosphere like the twilight of a night vision.' This vision is, Hardy insists, 'for one person at least'—his transfigurations are always attended with caution: here the 'dark purple cast' only *seems* to exhale from the cliffs. The mystery lies as much in the beholder as in the beheld. The point is underlined with reductive humour in *The Woodlanders* (1887). John South has a terror of a particular elm tree that overhangs his house; he feels that it is menacing his life. As his neighbour tells the doctor from the town, 'The shape of it seems to haunt him like an evil spirit. He says that it is exactly his own age, that it has got human sense, and sprouted up when he was born on purpose to rule him, and keep him as its slave' (ch. 14). The rationalistic doctor decides that the tree must be cut down; but when the sick man awakes next morning to find it gone, he dies of shock. 'Damned if my remedy hasn't killed him!' is the doctor's comment.

The episode is characteristic of its author—the quizzical curiosity with which he regards the superstition, the seriousness with which he takes it ('Others have been like it afore in Hintock') and the humour at the rationalist's expense, even the fact that the episode has its economic consequences. All Hardy's novels show this fusion of inner and outer worlds of experience which is the nearest he gets to the presence of the supernatural. He absorbed the preternaturalistic traditions of his rustic background, but not on their own terms. 'Half my time—particularly when writing verse—I "believe" (in the modern sense of the word) . . . in spectres, mysterious voices, intuitions, omens, dreams, haunted places etc etc. But I do not believe in them in the old sense of the word any more for that . . .'.[4] Behind Hardy's statement lies the distinction between the preternatural and the paranormal. As he wrote to another correspondent,

My own interest lies largely in non-rationalistic subjects, since non-rationality seems, so far as one can perceive, to be the principle of the Universe. By which I do not mean foolishness, but rather a principle for which there is no exact name, lying at the indifference point between rationality and irrationality.[5]

For all their refusal of religious comfort, and their scepticism as to supernatural realities, his novels resist any reductively materialistic attitude to human life.

To this extent he is at one with his contemporary, Richard Jefferies, in whom a close study of the life of field and farmland is combined with a mystical feeling for the energies of natural forces which proceeds well beyond anything that Hardy's cautious agnosticism would allow. As a young man, Jefferies would lie on the ramparts of the prehistoric earthwork of Liddington in Wiltshire, among 'the slender grass, the crumble of dry chalk earth, the thyme flower, breathing the earth-encircling air', and experience that sense of agelessness with which he begins his spiritual testament: 'Now is eternity; now is the immortal life. Here this moment, by this tumulus, on earth, now; I exist in it . . . To the soul there is no past and no future; all is and will be ever, in now.'[6] For Jefferies, the experience was born of an imaginative immersion in that life of vegetation and elemental processes which he was among the first English writers to isolate and record. *The Story of My Heart* (1883) is a rhapsodic work, full of a feeling of frustration and strain. Dismissing all human intellectual striving for a knowledge of God, Jefferies views humanity as trapped and distracted by its own scientific achievements and powers of intellection. Behind everything lies mystery.

The aged caves of India, who shall tell when they were sculptured? Far back when the sun was burning, burning in the sky as now in untold precedent time. Is there any meaning in those ancient caves? The indistinguishable noise not to be resolved, born of the human struggle, mocks in answer.[7]

Had E. M. Forster come across that passage in his youth?

There is nothing preternatural about Jefferies's experience: what he yearns for is the genuinely supernatural as expressed in material terms, a state of transcendent ecstasy that was to be evoked more slackly in a good deal of popular 'nature mysticism' by later writers, as well as in others of a more thoughtful (though no less impassioned) kind, such as Mary Webb, Margiad Evans, and, more didactically, Henry Williamson, an avowed follower and advocate of Jefferies. This attitude to nature represents a fusion of the Romantic poets' response to landscape as being a reflection of, even a formative force upon, human emotions, with the belief that such a conviction was itself life-enhancing, humanity's age-old inheritance without which it was spiritually dead. This creed is modified in other writers by a realization that the forces of materialism are not to be checked or refuted by any considerations of this kind. The tension between the will to believe and the frustration of that belief by a reluctant rationality is an important ingredient both in Hardy's novels and in those of E. M. Forster and D. H. Lawrence also.

In Forster's *The Longest Journey* (1907), for example, the Wiltshire landscape is invested with emotional significance substantiated by precise geographical analysis: something momentous is hinted at in the evocation of place. Rickie Elliot's moment of testing, when he is told that he has a bastard brother, happens in a prehistoric sacred site, and his movement into the successive grassy ramparts corresponds to a movement of ritual initiation; the nearby Cadover, home of the witch-like Mrs Failing, is referred to as 'the perilous house'. Similarly, the Dell near Cambridge, where Rickie begins his ill-starred relationship with the materialistic Agnes, is invested with numinous properties: Agnes calls out to him like a wood nymph, and yet Forster makes it clear that this mythic quality is really the product of Rickie's imaginative perceptions. Sympathetic to belief in a supernatural order, he was unable through his own rationality fully to endorse such an attitude. In his early novels and tales he plays, sometimes wistfully, at other times sardonically, upon the psychic borderland.

Some of these tales, notably 'The Story of a Panic',[8] deal with the disturbing incursions of the great god Pan. Here Forster is writing in a literary tradition that was already well established. But his elementals are less destructive than are those of Arthur Machen or Algernon Blackwood, and their incursions are recounted with a good deal of playful and at times sarcastic humour. While prepared to be whimsical about whimsy, Forster is well aware of the emotional and imaginative demands of human sexuality, evasion of which produces the whimsy. His stories are thus comparable less with those of the visionary Arthur Machen, than with those of the sophisticated Saki. Whereas the latter, in such tales as 'Gabriel-Ernest'[9] and 'The Music on the Hill',[10] makes

use of preternaturalist motifs to score satirical, at times vindic-
tive, points, Forster employs them as means of propagating his
deeply felt humanistic beliefs. He subscribes to the spirit of religion, while baulking at the absolutes of dogma. Indeed, he sees
the two as being radically opposed: in his early fictions, a priest
is never the recipient of a theophany.

Lawrence, likewise, distrusted the professional clergy. In his
work, a strong emotional response to landscape is matched with
a sharp, and at times an angry, awareness of the damaging muddle
produced by a philosophy of literal-minded materialism: he looks
to a recognition of the energies of nature to replace traditional
religious pieties. Like Forster he was a frustrated supernaturalist.
At its best his writing can achieve a remarkable fusion of human
responses with ordinary sights and sounds, a celebrated instance
being the passage in *The Rainbow* (1915) in which Anna and Will
gather in the sheaves by moonlight. This novel is Lawrence's
most determined grasp at the absolute as embodied in the life
of the earth and that of the human beings in tune with it. The
rainbow itself combines the perfection of the arch with the
transitoriness of its components: Lawrence is prepared to recognize that, for human nature, knowledge of the absolute must
necessarily be relative.

Significantly, his few excursions into the ghost story are among
his least convincing creations. Where he does get close to a sense
of the conventionally numinous is in his response to the suggestiveness of landscape and to the religious associations of traditional sacred sites. In *Kangaroo* (1923) Richard Somers's reactions
to the prehistoric monuments of Cornwall represent his creator's
most systematic attempt to evoke the magical atmosphere readily aroused by places of this kind.

He had a passion, a profound nostalgia for the place . . . Old presences,
old awful presences round the black moor-edge, in the thick dusk, as
the sky of light was pushed upwards, away. Then an owl would fly and
hoot, and Richard lay with his soul departed back, back into the blood-
sacrificial pre-world, and sun-mystery, and the moon-power, and the
mistletoe on the tree, away from his own white world, his own white,
conscious day. (ch. 12)

But the sense of strain is obvious. In seeking to communicate the
incommunicable, Lawrence is forced to coin words and phrases,
and while resorting to language with a pseudo-scientific resonance,
to fall back on imagery out of hackneyed occultist literary stock.

His definitive supernaturalist work is the novella, *St Mawr* (1925). In it there is an interesting discussion of the nature of Pan. The comparisons with the interpretations of Machen and Forster are telling.

[Pan] was the God that is hidden in everything. In those days you saw the thing, you never saw the god in it . . . If you ever saw the god instead of the thing, you died . . . Pan was the hidden mystery—the hidden cause. That's how it was a Great God. Pan wasn't *he* at all: not even a great god. He was Pan. All: what you see when you see in full. In the daytime you see the thing. But if your third eye is open, which sees only the things that can't be seen, you may see Pan within the thing, hidden: you may see with your third eye, which is darkness.

That concept of the third eye is more helpful than are the earlier graspings after a vocabulary of earth-forces.

In Lawrence's world metaphysical concepts are actualized in physiological terms: one finds an affirmation of what may be termed the hyper-natural at the expense of the preternatural.[11] His work is full of restless aspiration after an experience of life which the materialistic spirituality of his contemporaries fails to satisfy, indeed which it degrades; and in the beautiful passage in which a Welsh groom speaks about the indigenous beliefs of his childhood world, Lawrence describes a magical mental cosmos which stylistically resembles recreated folklore. But by the end of *St Mawr* everything dissolves. The supernatural is not to be attained through the immanent divinity of Pan, either through imaginative beliefs or through sexual experience: both have been contaminated. The novella concludes in the New Mexican desert, an image of the total otherness of the absolute; and Lawrence safeguards that otherness by making even this climactic vision the experience not of his protagonist (which could have the effect of 'closing' the story, of purporting to clinch an argument), but of another woman, who has both responded to the beauty and been driven to despair by the hardships involved in living with it. In these closing passages the intimacies and well-nigh sexual yearnings of Jefferies are displaced by an awareness of transcendence that remains implacably aloof.

Countries of the Mind

To achieve this sense of otherness Lawrence himself needed to go abroad: the English landscape, however eloquently he responded to it, was claustrophobic to him. However, other writers, less ambitious in scope, were able convincingly to celebrate the presence of the mysterium in terms of a known, loved, and precisely studied local world. As Kenneth Clark observes, the English sense of beauty is usually located in 'a lake and mountain, perhaps a cottage garden . . . but, at all events, a landscape. Even those of us for whom these popular images of beauty have been cheapened by insensitive repetition, still look to nature as an unequalled source of consolation and joy.'[12] This sense of familiarity, this all but encoded system of responses and corresponding emotions, is itself the product of an artistic and literary orthodoxy that stems from the Romantic sensibility. It provides an alphabet for the emotions, in the words of John Clare,

> A language that is ever green
> That feelings unto all impart,
> As hawthorn blossoms, soon as seen,
> Give May to every heart.
>
> ('Pastoral Poesy')

There is a precision and a rightness in that analysis which is all too easily sentimentalized: it is the novelists who, while endorsing the feeling, can portray it in its context and at the same time follow it through to its deeper implications, who may be called the genuine rural fictive supernaturalists. The first of these capacities can be seen in a writer like Constance Holme (1880–1955), in whose work is found 'a significant tension between the mode she describes as "the green gates of vision"—an authentic but specialised survival of the green language of Clare—and a rather sharp, placing, informed observation of people and events . . .'.[13] As Raymond Williams here indicates, Holme writes of her native Westmorland through the eyes and hearts of those who work and perceive and love it. She is a good instance of a novelist concerned with the impact of a particular regional landscape upon its inhabitants' subjective worlds and capability for visionary experience. In no less than six of her eight novels the protagonists are given over, heart and mind, to a particular place, through which they attain an inner life transcending space and time. But there is nothing sentimental about Holme's understanding of the cost of such an experience, however lulling her prose cadences can sometimes be. The power of place is bound up with the power of custom, space with time; and the retention of the one is conditioned by acceptance of the inexorable encroachments of the other. The pain involved in such a dual awareness is captured in *Beautiful End* (1918), a poignant account of the consolations and demands of the transfiguring imagination as they affect an elderly retired farmer. Unlike the majority of novelists who deal in the sensibilities, Holme does not confine her accounts of such experiences to the articulate educated classes.

Her fictive landscape is an actual one, the marshes and surrounding fells of the Kent estuary at the head of Morecambe Bay. Its geographical formation amounts to a natural metaphor, with the sea the recurrent threat to the security of the land, a reminder of the forces of contingency and spiritual expansiveness which perpetually challenge the urge to static containment and the aquiescence in recurring cultural patterns. Such symbolism governs all Holme's novels, but especially the earlier of the two supernaturalist ones, *The Old Road from Spain* (1915), which externalizes it by drawing on local legends that were part of its author's own ancestral traditions. The Huddlestons of Thorns Hall possess a hereditary wandering streak, disturbing and destructive to the more conservative members of the family, who are wedded to the land they own. The 'Spanish' members result from the adoption of a shipwrecked survivor of the Armada. His curse upon the English involves a flock of Herdwick sheep washed ashore at the same time: each head of the family is to know of his death in advance, when the sheep come down from the fells into the park to warn him. Holme unfolds the legend piecemeal, as Luis, a 'Spanish' Huddleston invalided home from the diplomatic service, endures his own antipathy to the austere northern landscape, watches his elder brother's terror as the sheep invade the park, and finally undertakes an expiatory voyage in the hope of lifting the curse. He succeeds, at the cost of his own life. As in other of the author's novels, the fulfilment of a dream is dependent on that dream's destruction.

The behaviour of the sheep is most convincingly rendered, and is plausible not only because of its acceptance locally or because, as Holme remarks, 'Their conservatism . . . amounts to a passion almost terrifying in its intensity',[14] but because there is nothing overtly preternatural about it. Save for their silence when they leave the fell, the sheep act as sheep do. As a result, the supernatural is presented as the fulfilment of the natural, and as the revelation of a dimension that is usually unperceived. It is therefore a failure of authorial tact that a motor car should break down when the sheep approach it—not so much because the car is thereby denatured, as because the sheep are. Their actions are not preternatural.

The Old Road from Spain evokes an authentic sense of the mysterium. In purely temporal terms, the story of the curse and its undoing outrages all notions of justice; but considered metaphysically as evidence of timeless laws, it satisfies through its imaginative logic. The book is the portrait of a troubled consciousness, and shows that this most localized of novelists could debate her own emotional preferences. The 'green language' is concerned with voyaging abroad in mind, not merely with cultivating a domesticated landscape.

Where the latter is the case, as in Holme's final novel, *He-Who-Came?* (1930), the result is parochial: the book reads more like a fairy-tale than a novel about the supernatural. None the less, it carries a certain sting. The young people's disowning of the white witch, Martha, when her gifts prove more powerful than they had supposed, is tartly satirical as to the nature of the popular credence given to supernatural happenings; while Martha's relations with their omnicompetent mother slyly question the status of the woman who so satisfactorily fulfils the maternal and domestic roles prescribed for her. Martha, neither wife nor mother, diverts her creativity into witchcraft, albeit of a benign variety. But the distinction between black and white magic becomes blurred; in the end it is the farm and the country round it which exude a power encompassing them both. Once again the naturalistic overcomes the awareness of the preternatural.

A more sure balance between the two is maintained in the novels of Constance Holme's contemporary, the Shropshire novelist and poet, Mary Webb (1881-1927). It is instructive to compare her account of the rocky outcrop known as 'The Devil's Chair' with that given by Lawrence in *St Mawr*. The latter's account is terse and cryptically evocative:

the knot of pale granite . . . was one of those places where the spirit of aboriginal England still lingers, the old savage England, whose last blood flows still in a few Englishmen, Welshmen, Cornishmen. The rocks, whitish with weather of all the ages, jutted against the blue August sky, heavy with age-moulded roundnesses.

Webb, on the other hand, in *The Golden Arrow* (1915) spells out this definition in terms of the imagination of the local people.

For miles around, in the plains, the valleys, the mountain dwellings it was feared. It drew the thunder, people said. Storms broke round it suddenly out of a clear sky; it seemed almost as if it created storm. No one cared to cross the range near it after dark—when the black grouse laughed sardonically and the cry of a passing curlew shivered like broken glass. The sheep that inhabited these hills would, so the shepherds said, cluster suddenly and stampede for no reason, if they had grazed too near it in the night. So the throne stood—black, massive, untenanted, yet with a well-worn air. It had the look of a chair from which the occupant had just risen, to which he will shortly return. (ch. 4)

The carefully built-up account of observation, superstition, and hearsay evidence justifies the authorial flourish at the end, and evokes the numinous in a more convincing way than do Lawrence's textbook jottings about aboriginal blood.

Mary Webb's tales of shepherds, farmers, and smallholders, notably *Gone to Earth* (1917) and *Precious Bane* (1924), are so steeped in folklore that they ask to be read as reflections of the power of a particular landscape upon its inhabitants' inner lives: their physical settings have compulsive, almost authorial properties. In Webb's philosophy, however, any superstitious belief in the preternatural is irrelevant to an apprehension of the genuinely supernatural, which is only to be attained through the capacity for love, both of human beings and for the animal and vegetable creation. There is much in common here with the Christian outlook of George MacDonald. Webb's novels are the product of a reading of life that is shaped by natural forces and the play of weather and of local legends, traditions and customs: the individual psyche is part of a cohesive, sometimes cruel and threatening spiritual world, in which a belief in the benevolent power of the unseen is both an inspiration and a necessary form of self-protection.

In *The Golden Arrow*, for instance, it is the protagonist's loss of religious faith which leaves him vulnerable to the terrors of primitive superstition. Webb's belief in a spiritual order inherent both in nature and in the human heart has for its corollary the contention that a materialistic philosophy forfeits the power to draw comfort from natural beauty.

To the gayest heart, the most securely grounded in optimism, fog is deadening, and on a solitary mountain-side it is nerve-wracking. To Stephen, with his desperate need of joy, his craving for immortality and his agonised conviction that no such thing existed; with his desire for some one to worship and his grey certainty that there was no one; with his lust for clarity and certainty and his finding of gloom and stark negation—it was an unbearable witness to his worst fears. He wanted, as he could not get free of his desperate spiritual ill, opiates of some kind to deaden it. Here were no opiates—only reality... (ch. 45)

Webb's response to landscape is part of her religious vision; her Shropshire is interiorized in a way that Constance Holme's Westmorland is not. But both writers affirm a unity between the physical and the spiritual dimensions of human experience which is the antithesis of the materialistic philosophy of the post-Romantic era. From within the Wordsworthian tradition they write affirmatively, without self-questioning, and refuting that philosophy rather than engaging with it through the alienated attitudes set forth, however reluctantly, by Hardy, by Lawrence, and by Forster.

John Cowper Powys

One writer who does contrive to combine the romantic assertions of Holme and Webb with the expression of an analytic scepticism is John Cowper Powys (1872–1963), the supreme English master of indigenous imaginative supernaturalism. All his major novels are concerned with the nature of human perception and its response not only to the world of inanimate nature but to that of irrational or super-rational experience. These books form a succession of exploratory forays into the imaginative processes of self-liberation; and in them, the geological nature of particular landscapes, and the historical traditions of particular places, have a determining part to play. All the fictional characters are in a dynamic relationship with their environment.

In his first three novels, the landscapes of Somerset, East Anglia, and West Sussex, although described with a wealth of detail and psychic sensitivity unmatched since Hardy, remain essentially background. They are observed, encountered, discussed; but they are not integrated in any vital way with the characters who people them. But in *Ducdame* (1925) Powys ventures closer towards his real concern, employing in the process a whole range of effects that are avowedly designed to be regarded as preternatural. Several of these may be well-worn, but Powys makes them very much his own.

Thus the reluctance of the young Dorset squire Rook Ashover

to provide his ancient family with an heir is assailed by a tremendous psychic pressure emanating from the tombs of his ancestors. This pressure, the outcome of decades of tradition, on two occasions materializes as a cry. 'It was louder and more appalling than the cry of any wild creature . . . it seemed to come to him through some heavy, remote intervening substance . . . it suggested the united exultation of a host of people buried underground' (ch. 9). It is a noise recorded intuitively rather than with the ear; and Rook hears it as he is about to enter his cousin's bedroom on the night when she conceives their child.

He is also subjected to the forces of white and black magic, the former at the hands of an old gypsy who voices the confused images and beliefs of local folklore, and the latter through the hostility of the demented parish priest, obsessed with a nihilism that ultimately destroys him; but the preternatural elements are subsumed into the landscape. By a process of suggestion (called by G. Wilson Knight 'etherealising')[15] Powys, without in any way lessening the physical actuality of the woods and fields round Ashover, succeeds in evoking a supernatural hinterland; Rook describes it as 'the old Platonic idea of a universe composed of mind-stuff, of mind-forms, rarer and more beautiful than the visible world'. Powys's technique in eliciting such a feeling is thorough and very subtle. By the use of actual local names he summons up a mysterious past. Two fields sloping up to the fir-crowned Heron's ridge are known as Battlefield and Dorsal. The first name is obvious enough; but the second is mysterious, in the way that many field names are mysterious, and adds to the dream-like quality of the narrative.[16] Still more effective is the use of an antiquated signpost carrying the single word 'Gorm', a place that apparently no longer exists, if it ever did. The riddle is more suggestive of the preternatural than is the explanation (in rich Powysian Mummerset) that 'un were a girt devil's name, writ on thik board for to guide boggles and ghosties . . . This place bain't a place for neither thee nor me, mister!' (ch. 12).

Powys's suggestion of an other-world is at its most effective, however, in the remarkable scene where Rook, trudging down a lane across the water-meadows, encounters a young man on horseback who turns out to be the ghost of his unborn son. The experience is anticipated by his state of mind; he longs for oblivion, is riddled with self-hatred, and rejects the material daylight world. The landscape accordingly goes grey, in a manner consonant with heat haze; and out of it the horseman appears, with a feeling of complete naturalness. He is no malign spectre, rather a witness to the existence of that Platonic realm, or Cimmery Land as the gypsy calls it, which alone can give meaning to Rook's aching sense of emptiness.

Ducdame is meditative rather than dramatic. Using familiar romantic elements associated with supernaturalist fiction, it subsumes them into an impression of other dimensions encompassing the static lives of its characters. But there is only the most tenuous and fragmentary sense of the occult. Powys's deployment of the phraseology of supposition (even the opening sentence containing the words 'it might almost seem as if'), and his resort to fantastic and sometimes distracting similes, disturb one's sense of rational stability; but they are offset by the meticulous, not to say pedantic, records of weather and plant life which make *Ducdame* memorable more for its setting than its story. At once constrictingly local and persistently subjective, the novel none the less leaves an impression of 'a world without bounds, of a universe where man's consciousness is the germ of creation'.[17] In *Wolf Solent, A Glastonbury Romance,* and *Porius* Powys pushes this use of a 'green language' further still.

In *Wolf Solent* (1929), however, the naturalistic mode predominates. The whole thrust of the novel is towards a refutation of belief in the preternatural, and a throwing-off of the burden of credulity. This is, however, only part of a more general purpose, the evaluation of a man's private mythology and inner language of sign and symbol as guides to reality and self-fulfilment. The West Country landscape asserts its reality over the miasmas of fear and superstition. Nowhere does the preternatural make an appearance, even though its presence is half-suggested; had it done so, it would assuredly have been explained away with Radcliffean thoroughness. *Wolf Solent* is a novel about a man whose dreams come true—by which is meant a man whose wishes are fulfilled in such a way that their dream element is dissolved and the man himself wakes up into reality. For all its stressfulness and tension, the novel is at heart a comedy, not least because, while subverting its own preternaturalistic tendencies, it retains a liberating feeling of romance.

And *A Glastonbury Romance* (1932) is likewise comedic in intent; but here the comedy is at the expense of the reader rather than of the protagonist. For one thing, there are at least half a dozen protagonists; for another, Powys is deliberately undermining expectations as to any rationally determined materialistic plausibility. In the opening paragraph a train arrives twenty minutes *early*; elsewhere the sun is seen to rise in the west; moreover an entire cosmology of sun, moon, and Good/Evil First Cause is confidently discussed as though such matters were generally received as being appropriate matters for debate. The physical setting is depicted in scrupulously naturalistic detail, but no less authoritative descriptions are given of the unconscious world of dreams and fantasies and psychic emanations. For this novel is an outright assertion that that world is the true source and motivating power behind the world of action, deliberation, and rationalized behaviour which is the naturalistic novelist's basic premiss and concern. Powys's Glastonbury is both a carefully portrayed community and a manifestation of the psycho-physical cosmos. The preternatural merges with the occult in the presentation of a supernatural multiverse.

The novel draws on the Glastonbury legends and on Arthurian imagery, Powys himself declaring that the book's true heroine is the Grail, whose 'essential nature . . . is only the nature of a symbol. It refers us to things beyond itself and to things beyond words.' The book's message is that 'no one Receptacle of Life and no one Fountain of Life poured into that Receptacle can contain or explain what the world offers us'.[18] Leisurely, humorous, lyrical, dramatic, A Glastonbury Romance celebrates a total relativity.

Powys's use of his supernaturalist motifs is as various as are the book's characters. There is the background of elemental powers; there are several cases of astral-projection, miraculous events, and a blend of occultism and natural magic; but as one of Powys's most percipient commentators points out, 'the psychic and occult elements in A Glastonbury Romance (and elsewhere) are introduced in such a way that they in fact cannot be seen as "elements" but only as integral subunits of a higher order of understanding—higher than logic but also higher than occultism'.[19] Most of the principal characters witness epiphanies or preternatural manifestations. Some of these may be regarded as purely subjective. A case in point being the anguish of the guilt-ridden sadist Owen Evans as he hangs upon the cross during the Midsummer pageant; or John Geard's feelings of union with his Lord as he makes his solitary and unorthodox Easter communion in his wife's front garden. Other moments are overtly preternatural, as in John Crow's vision of Arthur's sword at Pomparlés Bridge: Geard's encounter with Merlin in the haunted gallery at Mark's Court; and, more fully supernatural, Mary Crow's and Sam Dekker's separate and very different visions of the Grail. (It is typical of Powys's approach that these should be two people who strongly dislike and distrust each other.)

John Crow is a natural sceptic; yet when, staring at the de-composing body of a cat while in a mood of bitter self-disgust, he sees 'an object resembling a sword' flash past him and disappear into the river mud, he at once accepts the event as preternatural. The narrator attributes it to a conjunction of psychic forces and the peculiar powers of the legendary Arthur: Crow has, as it were, tuned into the prevailing mood of the place, the residue of the king's own mood when he threw away his sword. The event is, in occult terms, explicable; even, in spiritual terms, accidental.

But when the miracle-working evangelist John Geard en-counters Merlin, something more momentous is in progress. Geard is sleeping in King Mark's gallery as a result of a dare; it is a deliberate tryst. Powys makes due preparation for a ghostly visitation with descriptive details such as the terror of a servant, the stone walls of the haunted room, the howling of the wind—only to offset them by his circumstantial account of Geard's faintly grotesque preparations for sleep. Then Geard is awoken by a piercing cry almost within his own head: 'Nineue! Nineue!' This moment of alarm and anguish at the name of Merlin's treacherous companion is followed by the prosaic sound of a man and woman in the room below pissing into a chamber-pot. To forestall a second cry, the terrified Geard calls out 'Christ have mercy upon you!' and peace returns. From this moment on he takes up his vocation to play the role of Merlin, finding a temporary Nineue in the nocturnal visit to him of his host's youthful daughter, Rachel.

In this extended episode Powys achieves a fusion of the homely with the weird, of refined subtlety with gross earthiness, designed, as is the whole of A Glastonbury Romance, to accommodate conflicting aspects of human experience of the unseen. The result here is extraordinarily complex. Geard alone hears Merlin's cry, and interprets it as one of need and longing uttered in a place where, according to legend, the magician reduced King Mark to dust. Geard is a man who renounces power; he never exploits his thaumaturgic gifts for personal ends, and it is his consciousness which understands the phenomena for what they are. Rachel, on the other hand, hears the cry miraculously, from an impossible distance. Through her, the authenticity of the supernatural nature of the incident is confirmed.

At the same time, the couple with the chamber-pot prevent the scene from being rarefied; and Powys is to relate the visionary and the excremental still more effectively in Sam Dekker's vision of the Grail. Whereas Mary Crow receives her epiphany in the joyous perception of dawn light upon the Abbey ruins, Sam receives his at dusk, seated on a barge moored to an ancient post upon a river bank. The initial aesthetic delight aroused by the sight of this post (a characteristically Powysian moment) leads to a complex meditation in which Sam comes to have an overwhelming sense of his oneness with inorganic matter. This reminds him of a dead fish in his father's aquarium and, as he sits with his head turned towards the three holy hills of Glastonbury, of the world-fish, Ichthus. The vision of the Grail itself is preceded by a crashing pain and the feeling of a sharp object being thrust up his anus: he sees 'a globular chalice that had two circular handles. The substance it was made of was clearer than crystal; and within it there was dark water streaked with blood, and within the water was a shining fish' (ch. 28). Sam's version of the Grail achiever's question is to ask 'Is it a tench?' A keen naturalist, his own instinctive interest dictates this response. (And any other novelist would have made the dead fish in the aquarium a tench indeed—but not Powys: it was a minnow.) Next day Sam gives an enema to an old man suffering from constipation; the ungainly experience thus parodies Sam's own before he sees the Grail. Equally, Sam's spiritual experience is described in terms appropriate to an enema. A total imagistic blending of the physical and the spiritual is achieved. The seriousness is a matter of realism, not of solemnity.

Where plot is concerned, ramifications of these experiences are negative. A Glastonbury Romance is not a novel which uses the preternatural to make moral or dramatic points. But the manner in which the author portrays these preternatural experiences, the stress he lays on the mental and physical conditioning of their recipients, results in an overall view of life as itself supernatural. The Grail is nowhere and everywhere, and like the Mother Goddess Cybele evoked at the conclusion, Never or Always.

A later novel, Maiden Castle (1936),[20] dramatizes the clash

between those who possess a supernaturalist, personally mythological view of life and those with whom they have to live. The
existence of a metaphysical spiritual world becomes a source of
contention and is subordinated to the passionate life-illusion of
a thaumaturgic figure who, like John Geard, is presented by his
creator with a mixture of serious curiosity and compassionate
humour. The novel plays on the imaginative consequences of
archaeological research, and its preternatural and visionary components are barely developed in metaphysical terms. In *Owen
Glendower* (1940) likewise, the abnormal happenings and psychic
manifestations are subordinate to the human drama; indeed, Owen
deliberately renounces his psychic powers before going into battle.
But although subordinate, they do in fact colour the narrative,
and provide instances of the overall insistence on the supremacy
of imaginative vision which makes this novel the most purpose-
ful of Powys's fictive writings on this theme, one which ends on
a magnificently dying fall as the defeated prince takes refuge in
an ancient hill-fort, to be in his turn supernaturalized as a portion
of the Welsh people's mythological understanding of their history.

And in *Porius: A Romance of the Dark Ages* (1951)[21] Powys
produces his own fusion of history and myth; it is a novel pervaded by a sense of numinous space as much as *Owen Glendower*
is one pervaded by a sense of numinous time. Set in North Wales
at the end of the fifth century, its intricately structured narrative
is built around the figure of Merlin (here called Myrrdin Wyllt).
In this extraordinary book the natural and the supernatural form
a single common consciousness, the magical and the mysterious,
the comic and the homely, the tragic and the pathetic becoming
aspects of each other. Myrrdin Wyllt, like John Geard and Owen
Glendower, is an embodiment of a magic that foregoes the use
of power, a figure also who transcends the limitations of traditional Arthurian iconography. The fact that Powys chooses a single
week in a virtually unknown historical period frees him to set in
motion ideas known to be current at the time, but without having
to ascribe them to previously documented characters: simultaneously his close study of racial and cultural differences keeps his
novel from drifting into fantasy. Only one overtly supernatural
event occurs, when Myrrdin Wyllt transforms the owl Blodeuwedd
back into a woman: this Arthurian romance has more to do with
the *Mabinogion* than it does with Malory. However, the atmosphere is so remote and strange, so steeped in a world of primeval

forest, heath, and swamp, that there is no shock of the preternatural about the episode. The romantic and the naturalistic
modes are in complete accord, and Myrrdin's action can be seen
as a restoration and fulfilment rather than as an interference
with nature or a case of the arbitrary manipulation of abnormal
powers. *Porius* is an achievement of a unique kind; in comparison
even *A Glastonbury Romance* appears conventional. It displays
the fruition of an understanding of the supernatural that is entirely free from the materialism of dogmatic presentation and
systematic ordering. Powys's philosophy is one in which everything that can be imagined is a valid aspect of reality; and *Porius*
is a dramatized rendering of such a point of view. However, in
his subsequent novels the eccentric and the marvellous tend to
swamp the naturalistic elements, while the wayward and endlessly
self-contradictory stories of space-travel which reflect his final
speculations show his work not so much crumbling to a halt as
evaporating into intellectual mist.

Powys's attitude amounts to a considered refutation of traditional materialistic assumptions as to what are the priorities in
human life, and an insistence upon the importance of day-dream,
reverie, and what is usually designated the life of the subconscious
over that of the purely rational and cerebral aspects of personality. (In this respect he is as much a modernist as are Lawrence,
Joyce, and Woolf.) His novels are, in his own words, 'Simply so
much propaganda, as effective as I can make it, for my philosophy
of life. It is the prophecy and poetry of an organism that feels
itself in possession of certain magical secrets that it enjoys com-
municating.'[22] He goes on to add that 'I certainly feel conscious
of conveying much more of the cubic solidity of my vision of
things in fiction than it is possible to do in any sort of non-
fiction'—a significant comment in relation to fictive naturalism.
Ironically, Powys's quest for a panacea for human troubles, in
that fusion of physical with spiritual which is the supernatural,
is most satisfactorily presented in the least supernaturalistic of
the mature novels, *Weymouth Sands* (1934), which being concerned with the existence of a town in the consciousness, dreams,
and associations of its inhabitants, is in some ways the most
humane and accessible of them. But whereas in Forster and in
Lawrence this fusion of physical with spiritual manifests itself in
their corrective insistence on the need for a more instinctive and
free acceptance of sexuality, Powys explores the *nature* of the

division in individual consciousness, and makes a central issue of the abnormal psychic, sexual, and spiritualistic elements which are its outcome. In place of the alternating analyses and manifestos in Lawrence's work we have a presentation of extravagances, contradictions, and absurdities that questions the very basis of the orthodox concept of what is natural. In Powys's fiction 'natural' and 'supernatural' are not mutually exclusive terms. His world is one in which

objects are neither things sensually enjoyed by self-refining hedonism, nor mere symbols representing an ideal overworld fundamentally divorced from their objective actuality . . . he gives intensity to the act of perception by suggesting that particular state in which we all, at one time or another, have glimpsed the divineness of Being in the mysterious factualness of ordinary things.[23]

Powys's novels, like those of Lawrence, express a yearning to be at home on the planet. While not going so far as Blackwood's Irishman in *The Centaur*, they do portray the earth, if not as sentient being, then as a mother who cherishes her children while seeming to ignore them and neglect them, and who can be known in relationship and not merely as the inert object of incestuous rape. An acceptance of the objective existence of matter does not necessarily imply a materialistic attitude towards it. 'As a twentieth century novelist deeply and personally concerned with the notion of psychological dislocation, Powys [suggests] . . . that access to knowledge and self-enlightenment comes from nowhere if it does not come from a rooted, specific place in the real world.'[24] There is no way, in his writings, of reaching the supernatural except through an acceptance of the natural order.

If in order to evoke its strangeness the literary realization of the preternatural has depended on a close fidelity to physical realities, the attempt to convey a sense of the numinosity of landscape would seem to labour at a disadvantage. For here we are not concerned with invasive or disturbing forces, but with indwelling and all-encompassing ones, not with the strange but with the familiar. Only the sense of an immanent significance can make this kind of literature persuasive: otherwise one is left with the predicament analysed by Coleridge in *Anima Poetae*: 'In looking at objects of Nature while I am thinking . . . I seem rather to be seeking, as it were *asking* for, a symbolical language for something within me that already and for ever exists, than observing anything new.'[25]

Such a dichotomy between natural and supernatural as that expressed in the very language of novels like *The Lancashire Witches*, in which two kinds of fictive tradition exist uneasily side by side, is only partly bridged in the writings of nineteenth- and early twentieth-century novelists. A self-consciousness obtrudes, a temptation solipsistically to absorb landscape into the psyche: the fictive dramatizing of the kind of experience autobiographically described by Richard Jefferies is a difficult undertaking, witness the novels of Mary Webb and Constance Holme, both of whom can lapse into sentimentality when they lay stress too insistently on individual emotion. Again, the rather self-conscious voicing of estrangement and a yearning after reunion with the forces of nature, expressed in their very different ways by Lawrence and by Forster, is uneasy with itself. The numinosity sensed or sought-after seems only partially a matter of belief.

The eruptions of Pan, playful on the whole in Forster's fiction, are tensely contained in that of Lawrence—it is as if the puritan in him is defensively aware of the kind of collapse and disintegration portrayed by Arthur Machen: the latter may depict his native landscape of Gwent in a visionary light, but it is as a place of dangerous occult power as well as of healing beauty. Even Machen's most morbid and sensationalist writing quivers with a sense of this natural numinosity. But Lawrence, without having this particular sensibility, does provide a sense of the potential for such a religious view of material reality: several of his later novels and stories, even in their evocation of emptiness, leave one with a feeling of a place once spiritually inhabited and now vacated.

In this respect it is Hardy rather than Lawrence who anticipates the more substantial apprehensions that one finds in the work of John Cowper Powys—who acknowledged Hardy as his master. The elaborate descriptions of the Dorset landscape which characterize certain of the latter's novels are bound up with an invitation to contemplate and analyse. The accounts of Egdon Heath or the fields of Talbothays depend for their effect on the interaction between observer and observed: mystery is not confronted head-on, it is encountered and then penetrated. Thus, in *Far from the Madding Crowd* (1874), Norcombe Hill is said to be 'one of the spots which suggest to a passer-by that he is in the presence of a shape approaching the indestructible as nearly as any to be found on earth' (ch. 6). The rhetoric is muted by qualifications: 'suggest', 'approaching', 'nearly', the indefinite article governing 'passer-by'. On the other hand, Hardy chooses to write one of *the* spots, rather than the more usual *those*—the difference is in favour of greater certainty and an authoritative precision. Mystery is not simply asserted, it is substantiated.

One may compare this with Powys's account in *Wolf Solent* of two lovers walking in the fields.

Over this cold surface they moved hand in hand, between the unfallen mist of rain in the sky and the diffused mist of rain in the grass, until the man began to feel that they two were left alone alive, of all the people of the earth—that they two, careless of past and future, protected from the very ghosts of the dead by these tutelary vapours, were moving forward, themselves like ghosts, to some vague imponderable sanctuary where none could disturb or trouble them! (ch. 7)

The language of supposition is in continuous tension with the eloquence embodying the substance of the state supposed: only the concluding exclamation mark offsets the process of transfiguration that is going on. But in both passages nothing specifically supernatural has been described; what has been actualized is the presence of the mysterium in the human imagination's power so to interpret its surroundings as to make of them potentially sacred focuses for self-forgetful worship.

In the lore of geomancy the physical constituents of particular places are read as the vocabulary for a spiritual realm. Neither Hardy nor Lawrence, nor even Powys, goes so far as this; but all three allow for, and speculate about, the capacity to conceive such readings of the basic realities of space. Outside of such belief, however, those writers who extol humanity's relation to the energies of the planet may be said to hyper-naturalize the dormant responses of the imagination, and to champion, as all these novelists did, a more liberated response to human social customs and patterns of behaviour—an ideal of spiritual liberty which accepts the earth itself as magical, as a manifestation of the supernatural, though in a pagan rather than in a Christian sense. For in a civilization that no longer lays stress on the historical dimension of the Christian myth that shaped it, the mystery of space overlaps that of time. In the work of writers such as Lawrence and Powys the modern world's alienation from its natural environment results in that environment taking on the invasive, hostile qualities of the preternatural; precisely as these novelists explore the hyper-naturalness of phenomena, so they render it more possible to see them as latent evidence of the presence of the supernatural. Both writers convey a sense of the sacred, immanent in the world they see around them. Of Powys in particular it can be said that he takes the concept of the supernatural and domesticates it as the servant of the imagination's need to withstand the onslaughts of a physical and mental universe which would otherwise enslave and degrade it. He locates the mysterium within the capacity to recognize its presence.

7

THE ENEMY WITHIN

And when you gaze long into an abyss the abyss also gazes into you.

(Friedrich Nietzsche, *Beyond Good and Evil*)

There is a typically unnerving tale by L. P. Hartley which describes the doing to death of a novelist by a malignant creature of his own invention. The story has a parabolic quality. Hartley's fiction persistently records the disquiet of the comfortably-off middle classes when subjected to the social upheavals of the mid-twentieth century; and many of his otherwise naturalistic novels —*The Boat* (1949) particularly—convey a feeling of menace quite in excess of the rational deployment of their plots. The majority of his short stories recount abnormal or preternatural happenings, situations of entrapment and embarrassment, with some frightening instances of self-projection. Yet Hartley rarely uses conventional supernaturalist effects: his is largely a daylight world in which matters can go disturbingly askew, in which control is lost and sensory disjunction concludes in the engulfing of its victim in a malevolent and lethal clutch. The affinities with Sheridan Le Fanu are obvious.

This particular story, 'W.S.'[1] (the initials seem ironic) also confronts its readers with their own particular responses to the conceptualizing of human evil, and with their tendency to project ideas and fantasies of their own assimilation upon the life surrounding them. Self-identification and self-projection are inseparable properties of all who respond to fictional narratives; and twentieth-century critical inquiries as to the nature of such responses expose a reciprocal relationship between author and reader that is especially potent when applied to supernaturalist fiction in its more subversive aspects. For the fear both of fear and of fear's occasion is perennial. At the conclusion of his introduction to *The Supernatural Omnibus* Montague Summers quotes the answer given by Mme du Deffand, the friend and correspondent of the author of *The Castle of Otranto*, to the enquiry as to whether she believed in ghosts. 'No, but I am afraid of them.' The reply is as typical of its own century as it would be of ours. But the nature of that dread has changed. It is our own fear of which we are now afraid, for the world we live in is, we realize, a world we have in part created.

James Hogg and Robert Louis Stevenson

The contention that the past is a foreign country where matters are conducted differently is to some extent refuted by the manner in which certain ways of responding to supernaturalist themes in fiction keep recurring. Alarm at social change, distress at the erosion of religious beliefs, and a longing for Arcadian simplicities which the juggernauts of commercial and industrial exploitation have obliterated, find metaphorical expression in novels that posit an invasive, controlling, or immanent supernatural cosmic power that would judge or rescue or replace the prevailing sense of physical and mental dislocation—those of Charles Williams are obvious examples. But the Protestant religious tradition, by siting that power within the individual conscience, could impose an excessive psychological burden and thus undermine still further the individual's sense of personal security; as a result it tended to fortify against all odds those who were prepared in self-defence to deny themselves any notion of veridical relativity. It is not surprising that supernaturalist fiction should concern itself with the dialogue within the self, the psychic conflict between demonic and angelic forces. One sees such a conflict incipient in a speculative novel like *Melmoth the Wanderer*; and it finds its first mature expression early in the nineteenth-century in one of the masterpieces of supernaturalist fiction, *The Private Memoirs and Confessions of a Justified Sinner* (1824) by James Hogg (1770–1855). The author was both knowledgeable concerning the folklore of the Scottish borders and perceptive as to the psychological compulsions that underlay such beliefs. He saw a long way into the dark places of the spirit.

A Justified Sinner is especially interesting for its portrayal of a whole variety of responses to the subject of the supernatural, its plot involving the question of its own literary verification and authenticity. The book falls into three sections—or into two sections and a subsection. The first is 'The Editor's Narrative'. This anonymously provides what is at first sight a well-substantiated account of the events leading up to the disappearance of Robert Wringhim Colwan, Laird of Dalcastle, following the issue of a warrant for his arrest on suspicion of murder. These events are frequently enigmatic, but amount to a mystery story that involves elements of doubtful paternity, diabolic possession, and extreme religious fanaticism. The tone is crisp and detached, the opening couched in a vein of broad comedy suggestive of the contemporary work of John Galt;[2] but this also serves unobtrusively to slant the narrative in a direction which the recorded happenings need not imply. The puzzling events are left unexplained, and it is for the reader to deduce their meaning, vouched for as they are by a number of independent but not necessarily reliable witnesses.

The second part of the book consists of the 'Memoirs' themselves, which have come into the editor's hands in a manner so far unexplained. They are the work of the missing laird, and rewrite the previous narrative from the point of view of an obsessed and eventually hallucinating imagination; Robert Colwan is an extreme instance of antinomianism, holding a belief that God's elect, being predestined to beatitude, are above the moral law. The infatuated bigot falls gradually into the power of a mysterious young man professing the same beliefs; it transpires that he is the Devil. The concept is rendered more subtle by the manner in which the stranger takes on the physical appearance of whatever person it is with whom Colwan is at that time preoccupied: he is the externalization of interior desires. Gradually Colwan is reduced to a state in which, no longer conscious of his own actions, his divided self is partly given up to the diabolic energies. He ends as a fugitive on the English border, working as a herdsman, haunted by poltergeist disturbances, and awaiting his doom at the hands of his tormentor, who insists that they must die together.

The final subsection recounts the gruesome way in which the manuscript reached the editor's hands; it is exhumed from a suicide's grave, attention to which has been drawn by 'Mr Hogg' in a (perfectly genuine) letter in *Blackwood's Magazine* for August 1823. The final verdict of the editor as to the manuscript is sceptical, the author being dismissed as either a knave or a madman; but the evidence of his own earlier narrative in part refutes him. Hogg's novel is thus laid open to the reader in a way that looks ahead to certain twentieth-century experiments in open-ended narrative.

As a tale of the supernatural it is full of delicately managed detail. The *doppelgänger* motif suggested by Robert Colwan's encounter with Gil-Martin, his diabolic opposite, is tightened into actual possession; and any sense that this is purely a subjective experience is offset by the various independent witnesses of the two young men walking in each other's company. The issue is even canvassed within the novel itself, in the episode where George Colwan confronts the spectral appearance of his half-brother on the summit of Arthur's Seat while that brother crouches behind him preparing to attack him and send him hurtling down the precipice below. What George sees could, as a friend suggests, be a mirage, but the naturalistic explanation only serves to render a preternaturalistic one more likely.

Similarly, in the comedy of a serving man's account of the Devil's preaching to the townsfolk of Auchtermuchty, we have both an ironic comment at his expense on what is actually happening to his master, and an injection into the tale, as it nears its denouement, of an element of that folk belief which conditions what people know and see, an element put to grimmer use in Robert's account of his terrifying last days on the Border farm. A further piece of running irony lies in Robert's half-belief that Gil-Martin's boasted princedom is in fact Russia, and that his friend is the by now legendary Tsar Peter the Great. Again and again uncertainty as to the 'real' truth, and scepticism concerning simplistic solutions, marks the presentation of the story, an uncertainty mirrored in the book's various narrative devices and questionings of its own authenticity. But uncertainty does not mean confusion, any more than mystery means muddle.

The novel's peculiar power results from the interlacing of sharp tangy humour with touches of the eerie that are all the more unnerving for the apposite precision with which they are planted. Walter Allen's contention that the book is primarily a religious satire in the manner of Burns's 'Holy Willie's Prayer'[3] can only be sustained by ignoring the preternatural happenings. The ironic exposure of the logical consequences of a strict doctrine of election is unfolded with Swiftian inexorability; but what could be a comedy, and indeed starts as such, soon turns into a record of systematic, if involuntary, self-destruction. By objectifying the drama, the author induces a realization that evil in the heart can erupt and pervade the world: there are more ways of calling up the Devil than by direct invocation. Hogg's vision thus fuses the natural and the diabolic in a way that communicates a sense not merely of the creepy and unnatural, but of the supernatural in the fullest sense.

His mastery of the interiorized supernaturalist tale was to be echoed not only by George MacDonald but also by Robert Louis Stevenson. Both the story 'Markheim'[4] and the novella *The Strange Case of Dr Jekyll and Mr Hyde* (1886) are psycho-dramas. The former is an instance of the type of fable perfected by Nathaniel Hawthorne. Markheim, having murdered an antique-dealer for his money, encounters a being whom he takes for the Devil; this courteous and charming personage offers him help in return for a complete submission to the powers of predeterminism. His arguments are rational and fatalistic; but if Markheim's love of good is weak, he has a hatred of evil, and declines the bargain, whereupon 'The features of the visitor began to undergo a wonderful and lovely change: they brightened and softened into a tender triumph, and, even as they brightened, faded and dislimned.' The graceful prose underwrites a meliorism quite at odds with Hogg's grim version of the pernicious effects of the Calvinistic perversion of the doctrine of predestination; but both writers are agreed (as is George MacDonald) in refuting it. In Stevenson's story the sovereignty of good is delicately and ironically maintained: this devil of Markheim's imagining is the Satan of the Book of Job, the tester of faith, the adversary of man who is God's servant and who chastises him into salvation.

In *Dr Jekyll and Mr Hyde* the fable is couched in the form of a tale of what Stevenson calls transcendental medicine. Henry Jekyll's ability to incarnate his rejected passions in the person of Hyde is a dramatized enactment of schizophrenia. It is also a satire on Victorian hypocrisy, a symbolizing of society's capacity to compartmentalize its behaviour, and even of the role of woman in it, either as angel or as whore. The novel's most interesting aspect is its exposure of the workings of Jekyll's mind. Initially he sees the creation of Hyde as an expression of the ability to coast along in a state of moral neutrality.

It was on the moral side, and in my own person, that I learned to recognize the thorough and primitive duality of man; I saw that, of the two natures that contended in the field of my consciousness, even if I could rightly be said to be either, it was only because I was radically both; and from an early date, even before the course of my scientific discoveries had begun to suggest the most naked possibility of such a miracle, I had learned to dwell with pleasure, as a beloved day-dream, on the thought of the separation of these elements. If each, I told myself, could but be housed in separate identities, life would be relieved of all that was unbearable; the unjust might go his way, delivered from the aspirations and remorse of his more upright twin; and the just could walk steadfastly and securely on his upward path, doing the good things in which he found his pleasure, and no longer exposed to disgrace and penitence by the hands of this extraneous evil. It was the curse of man-kind that these incongruous fagots were thus bound together—that in the agonized womb of consciousness, these polar twins should be con-tinuously struggling. ('Henry Jekyll's Full Statement of the Case')

This is a beautifully precise expression of moral irresponsibility. To see the struggle as a curse, is, of course, heresy to the nineteenth-century Protestant ethic; and the book attains its great-est imaginative power in Jekyll's account of his failure to keep the two selves apart. Significantly, however, the language describ-ing Mr Hyde falls back upon the vocabulary of dualism:

he . . . thought of Hyde, for all his energy of life, as of something not only hellish but inorganic. This was the shocking thing; that the slime of the pit seemed to utter cries and voices; that the amorphous dust gesticu-lated and sinned; that what was dead, and had no shape, should usurp the offices of life. (Ibid.)

This is the traditional rhetoric of horror, one that was to be employed exhaustively in writers of the 1890s and after. More perceptive and original is Stevenson's initial account of Hyde as 'troglodytic' and, because of Jekyll's instinctive virtue and moral strength, as the younger and smaller of the two. This is an inter-esting variant on Lytton's ever-youthful Margrave, and is sugges-tive of the essential immaturity, the infantilism, of evil.

Henry James

The fact that Stevenson's fictions internalize supernaturalist ex-perience relates them to the work of Henry James (1843–1916), who may be regarded as the father of twentieth-century writing of this sort. Few authors seem more aware of the capacities and terrors latent in the human mind. James's knowledge was inherited. A year after he was born, his father underwent an experience that presaged both a breakdown and an eventual renewal of his life.

The afternoon was chilly and there was a fire in the grate . . . Relaxed, his mind skirting a variety of thoughts, he suddenly experienced a day-nightmare. It seemed to him that there was an invisible shape squatting in the room, 'raying out from his fetid personality influences fatal to life.' A deathly presence thus unseen had stalked from his mind into the house.[5]

One thinks of the lurking demon monkey in Le Fanu's 'Green Tea'.

The experience inevitably overshadowed a close-knit family like the Jameses, and the younger Henry's writings are frequently haunted by an awareness of the terrifying abyss into which consciousness can fall. It is less surprising that he should have written tales about the supernatural than that he should have written so few of them—overtly, at any rate. But much of James's work hovers on the borders of the unseen, and the crises in his major novels are spiritual crises. The supernaturalist tales reflect a development in technique and insight discernible in the fiction as a whole.

A few of the stories are intended as diversions: 'A Romance of Certain Old Clothes',[6] for instance, is a carefully built-up piece of gruesomeness, in which the horrific climax serves as a moral fable in the manner of Nathaniel Hawthorne; and 'Sir Edmund Orme',[7] with its silent daylight ghost in the sophisticated fashionable setting of James's 'society' novels, has the same effect. Best and most original of the lighter tales is 'The Third Person',[8] in which the manifestation of the ghost to two elderly ladies is an occasion for delicate satire (a playfulness not usually associated with the mode). More serious is the strictly parabolical 'Owen Wingrave',[9] in which the young scion of a military family, even though he rejects its traditions, fights his battle against an ancestral ghost to prove his manhood. The tragic outcome is a measure of James's fatalism concerning the formative priorities of hereditary taboo.

One group of tales in their ambivalence is both more representative of their author's art and significant for the handling of the genre as a whole. 'The Turn of the Screw' is the most famous of these; but that perplexing novella yields up a still richer meaning when read in the context of 'The Friends of the Friends', 'The Real Right Thing', and 'The Jolly Corner'.

Common to all four stories is the element of hide-and-seek: at times it seems uncertain as to who is the haunter, who the haunted. A foretaste of this effect is to be found in 'The Ghostly Rental',[10] an early story which James never reprinted. It is a fanciful tale, told with all the (relatively) young author's debonair facility. The narrator comes across a shuttered house which he intuits to be haunted. He gradually learns that it is—by the ghost of the estranged daughter of an elderly army captain, who comes there each quarter-day to collect the rent off which he lives. The narrator accompanies him on the next occasion, sees the ghost standing at the head of the stairs, and retires in alarm. On the following quarter-day he has to go himself, on the dying captain's behalf—to discover that the ghost is a living woman who has chosen this masquerade as a way of looking after her recalcitrant father. At which point she sees the old man himself, at what, it transpires, is the moment of his death. A real ghost has been substituted for an unreal one.

It is an implausible tale, but James was to use more than one of its ingredients again. In 'The Way It Came', renamed 'The Friends of the Friends',[11] uncertainty as to the supernatural becomes the crux of a personal drama. A man and woman, both of whom have experienced visions of their fathers at the moment of death, are fortuitously prevented from encountering each other, despite the repeated efforts of their friends. Eventually the man's fiancée arranges a meeting, only to frustrate it at the last minute, as the result of a fit of jealousy. That same night the other woman dies; and, either in body or in spirit, pays a silent visit to the man she now can never meet. The fiancée maintains that this is a ghostly visitation, an easier supposition to bear than is his, that it was a physical encounter. Aware of a continuing relationship between the two, she comes to realize that her own interpretation is even more destructive than is his. The other-world has proved more powerful than the material one, since it imposes its own verdict upon it. The story, indeed, is double: the narrator's interpretation of what she has been told emerges as a determining reality of a potency equal to the specific manifestation which she knows of only at one remove (and the reader at two). In this story, the actual 'appearance' being offstage, the spiritual relationship between the three people involved is fully developed on its own terms. The supernatural here assists James to convey what would otherwise be the incommunicable.

In 'The Real Right Thing'[12] a great writer's widow and his biographer are prevented by his ghost from writing an official life. Felt initally as a helpful presence, sifting papers, shifting books, it materializes as 'Immense. But dim. Dark. Dreadful.' The story is conventional enough; but one notices yet again the controlling force of the unseen, and also an exchange of hauntings, for the biographers would in their turn be haunting the dead man. James's ghosts are not intrusive spectres from an alien world, but the master guardians and interpreters of this one.

The thread between seen and unseen is drawn tighter still in James's final and most moving tale of the uncanny, 'The Jolly Corner'.[13] Spencer Brydon, a returned expatriate, becomes obsessed with the self he would have become had he not gone to live in Europe. Wrapped in this concern with the past, he spends hours wandering about the old New York house in which he has been brought up, in a half-fanciful quest for his other self. The event begins as a 'quaint' analogy which

quite hauntingly remained with him, when he didn't indeed rather improve it by a still intenser form: that of his opening a door behind which he would have made sure of finding nothing, a door into a room shuttered and void, and yet so coming, with a great suppressed start, on some quite erect confronting presence, something planted in the middle of the place and facing him through the dusk.

Brydon regards his retention of the house as a charming perversity, 'my refusal to agree to a "deal" is just in the total absence of a reason . . . There are no reasons here *but* of dollars. Let us therefore have none whatever—not the ghost of one.' The quest becomes obsessional. 'People enough, first and last, had been in terror of apparitions, but who had ever before so turned the tables and become himself, in the apparitional world, an incalculable terror?' All James's powers of subtle suggestion are at work in the account of the gradual materializing of the spectre. First Brydon realizes that a door he had left open is now shut—thus enacting his own analogy, since it belongs to a room with no other access. Afraid to open the door, he descends the stairs, only to find waiting in the vestibule a figure that covers its face with its hands; James here repeats his presentation of the spectral figure in 'The Ghostly Rental'. But when the figure makes its appearance known, Brydon sees the face as 'evil, odious, blatant, vulgar': the diminution of the adjectives is telling. Here the *doppelgänger* is horribly other; but at the end of the story Brydon is compelled to recognize that he has indeed found what he was looking for, and in rejecting it has 'come to himself'.

James produces a variation on this theme, one in which protagonist and *doppelgänger* are fused, in his unfinished late novel, *The Sense of the Past*, begun in 1900 and resumed abortively fourteen years later. The inheritance of an old London house and an obsession with past time lead Ralph Pendrel to encounter what is an exact *doppelgänger*, and thereby to be drawn back into the world of 1820—the two versions of the self change place in time. Pendrel's own beliefs reflect those explored by many writers on supernaturalist themes.

There are particular places where things have happened, places enclosed and ordered and subject to the continuity of life mostly, that seem to put us into communication, and the spell is sometimes made to work by the imposition of hands, if it be patient enough, on an old subject or an old surface. (Book First)

Ralph's adventure involves the characteristic Jamesian reversal. 'He had had his idea of testing the house, and lo it was the house that by a turn of the tables had tested him.'[14]

In the early chapters the author reflects on the tradition of supernaturalist fiction as practised by his contemporaries.

Nights spent in peculiar houses were a favourite theme of the magazines, and he remembered tales about them that had been thought clever—only regretting now that he had not heard on the retreat of his fellow-occupants (for was not that always the indispensible stroke?) the terrified bang of the door. (Book Second)

The bang is, significantly, terrified, not terrifying: it is at one with the mind of the auditor. This is in keeping with James's own practice in 'The Turn of the Screw',[15] in which, according to the Preface to the New York Edition, his object was to produce something 'of a nature to arouse the dear old sacred terror'.

His success in this case is indisputable. Their governess's conviction that two small children are being corrupted by the spirits of a former governess, Miss Jessel, and her lover, the valet Peter Quint, amounts to one of the most widely discussed ghostly tales in English fiction. Its thematic ambiguity has led to disputes as to interpretation which a reading of James's other tales in this kind can elucidate. Certainly the supposition that the ghosts exist solely in the governess's mind is too simple: James himself is clear that for him the ghosts are 'real'. In view of his other stories one recognizes that what the governess sees is a past which, through coincidence of situation (her own unreciprocated infatuation with her employer), is reanimated, and with which the children's experience of Quint and Miss Jessel enables them to comply, if only in response to what they feel the governess herself requires of them. In this story the mysterium is experienced as darkness and horror. The governess reflects Miss Jessel—just as, for her, Peter Quint reflects a degrading sexuality. In endeavouring to fight the evil she imposes it upon the children. They have been corrupted to the extent, it would seem, of what they have heard—Miles is expelled from school for 'saying things', things that possibly Flora passes on to the housekeeper, Mrs Grose. Apart from the governess's suppositions, there is no evidence that they see what she sees. And the fact that it is the governess herself, rather than Quint or Miss Jessel, who seeks control over them, is indicated not only in Miles's final cry to her, 'you devil!', but in the moment of peculiarly Jamesian indirection when Mrs Grose is seen by the governess looking at her as she herself has just been looking at Peter Quint—yet another case of the human being and the ghost exchanging places. The story's horror lies in the way the governess is drawn into a pre-existent atmosphere of evil (attested by the recollections of Mrs Grose) so that in her very attempt to fight it she makes it worse. By encouraging a possibility, she actualizes it.

'The Turn of the Screw' is a mystery, not a puzzle. The ghosts are not simply to be explained away as manifestations of the governess's sexuality. The children may be as corrupt as she thinks they are: the possibility is part of the total vision of evil. But an alternative reading is written into the text. The goodwill of the governess is objectively attested (though that too is doubtful) in the prologue, where the preparation for her story is momentous. At the end Quint is referred to Miltonically as 'the hideous author of our woe', just as Mrs Grose's room is at one point described as being 'swept and garnished'[16]—and thus presumably ready for seven other devils worse than those who are to be expelled. 'The Turn of the Screw' is a picture of the nature and work of evil in far more than a local or particular sense. When at the start it is suggested that the tale will tell whether the governess was in love with her employer, the reply is 'The story *won't* tell . . . not in any literal, vulgar way.' It does indeed not 'tell'; but its methodology suggests and creates what James himself was determined that it should.

Only make the reader's general vision of evil intense enough . . . and his own experience, his own imagination, his own sympathy (with the children) and horror (of their false friends) will supply him quite sufficiently with all the particulars. Make him *think* the evil, make him think of it for himself, and you are released from weak specifications.[17]

The tale emerges from the reader's mind much as the appalling goblin came from that of James's father.

In considering James's presentation of the supernatural one is aware of the refinement of the preternatural into a manifestation of psychic forces that are, to use religious terminology, the properties of the soul: in this respect his work has affinities with that of Hogg. In 'The Turn of the Screw' the manifestations are confined to the consciousness of those involved with them. The specificity of the apparitions, however, suggests that if they are hallucinatory they are also malign.[18] Since it is the governess alone who sees the ghosts, and since it is on her word only that we know of them, the story remains, however horribly, within the territory of the psychic rather than of the preternatural. It is the way in which it is poised on the borderline between the two, the way in which the author leaves it to the reader to embody the evil, that makes 'The Turn of the Screw' the masterpiece of terror that, as James himself might say, it so wonderfully is.

'The Turn of the Screw' is an extreme case; but the border between James's supernaturalist tales and his naturalistic ones is not easily determined. 'The Beast in the Jungle',[19] for instance, a chilling account of atrophied emotion and self-dramatizing self-enclosure, may rest on a personal fantasy; that fantasy determines the action (or inaction) of the story. The beast refers one back to the affliction of James's father; but in this case there is no perceptible vastation. Similarly, in 'The Altar of the Dead'[20] a private obsession with the departed takes on an overpowering spiritual force. Graham Greene's assertion that the story is 'ridiculous', since it betrays a disregard for the Catholic rituals available for commemorating the dead, is beside the point.[21] James is concerned with the inner world of the psyche, not with the encompassing world of spirit.

Elizabeth Bowen

In a preface written for one of Cynthia Asquith's *Ghost Book* anthologies, Elizabeth Bowen (1901–1973), wondering why such stories should still be popular, suggests that 'it may be that, deadened by information, we are glad of these awful, intent and nameless beings as to whom no information is to be had. Our irrational, darker selves demand familiars.'[22] In its sophisticated way the speculation is characteristically self-punishing: Bowen is a novelist whose surface delights of style and social comedy are sustained by a bleak awareness of the self-deception of which human beings are capable, and of their readiness to resort to fantasy and protective irony to keep them from penetrating the gulf that lies within. Hers is a defensive world, and as such representative of the mid-twentieth-century, with its reluctance to endorse wholeheartedly the outward manifestation of spiritual impressions: no wonder that her forays into supernaturalist fiction should be astringent, even at times satirical. For her, the genre is not so much parabolic as metaphorical—though a metaphor that may turn out to be more exact than one would wish to think.

As a novelist Bowen has strong affinities with James: indeed, if anyone is to be singled out as his successor, it is she. Both writers were meticulous practitioners of their craft; both were concerned with the finer points of sensibility and obsessed with the precarious condition of innocence and virtue in a corrupt world. Both possessed an idiosyncratic style that tended to condition its readers' responses; and in their shorter fiction both writers resorted to the preternatural. In Bowen's case a sense of the unseen hovers about a number of her novels also, to the extent that the imaginations of her characters interact with their surroundings with at times hallucinatory effect—*A World of Love* (1955) and *The Heat of the Day* (1949) are outstanding in this respect—but her work has even less contact than has that of James with the Gothic tradition of the past.

She does, however, exhibit a relish for the gruesome, and contributed two mordantly nasty pieces to Cynthia Asquith's anthologies; in the above-quoted preface to the second of these she remarks of ghosts that 'the guilt-complex is their especial friend'. The comment is appropriate to her own 'Hand in Glove',[23] a variant on James's 'A Romance of Certain Old Clothes'. More characteristic is the title story of *The Cat Jumps* (1934), a rare blending of satire with an extreme of psychic terror. A rationalist couple buy a villa in which an atrocious murder has lately taken place; and their weekend house-warming is assailed by hauntings and by their own superstitious terrors. The story is an apt comment on twentieth-century materialistic attitudes to the supernatural, and is both comical and eerie; it concerns, and tends to induce, hysteria. A refinement of such attitudes is to be found in 'The Apple Tree'.[24] A girl's remorse for the suicide of a neglected schoolfriend, whose body she has found hanging from an apple tree, emanates at certain seasons in the form of the tree itself: the repressed knowledge takes on tangible form, the spiritual element becomes incarnate. Ironically, the girl is exorcised through the efforts of a determinedly materialistic society woman straight out of the later fiction of Henry James. The preternatural happening is outfaced, pushed back into the dimension where it belongs.

This resolution is in keeping with Bowen's avowed preference for describing life 'with the lid on'. Such hypertension between different levels of reality is sustained in the collection of wartime stories, *The Demon Lover* (1947). In it the preternatural, though it surfaces in a variety of ways, is pervasive but latent. The destruction in wartime London and the consequent social fragmentations are presented (in 'In the Square') with surreal effect, an impression furthered by the author's angular and frequently eccentric prose:

the square looked mysterious: it was completely empty, and a whitish reflection, ghost of the glare of midday, came from the pale-coloured facades on its four sides and seemed to brim it up to the top. The grass was parched in the middle; its shaved surface was paid for by people who had gone. The sun, now too low to enter normally, was able to enter brilliantly at a point where three of the houses had been bombed away; two or three of the may trees, dark with summer, caught on their tops the illicit gold. Each side of the breach, exposed wallpapers were exaggerated into viridians, yellows and corals that they had probably never been. Elsewhere, the painted front doors under the balconies and at the tops of steps not whitened for some time stood out in the deadness of colour with light off it. Most of the glassless windows were shuttered or boarded up, but some framed hollow inside dark.

What might be a conventionally Gothic account of empty deserted houses is here rendered in terms of the dislocated consciousness of the perceiver: it is as though the nature of things has been turned inside out, like the bombed houses themselves. Except the first one, each sentence contains a negative, either as statement or as concept.

This ghostly terrain is peopled by a variety of psychic manifestations. In the unnervingly irrational title story a middle-aged woman is abducted by a dead lover—whose death in the First World War is, however, open to question (a materialistic explanation is just possible); in 'Pink May' an unseen ghost may after all be the conscience of the commonplace narrator; in 'The Happy Autumn Fields', most moving of these stories, a woman in a bombed house is transported in spirit into a nineteenth-century world through rummaging in an old box of family papers. Other stories, such as 'The Cheery Soul', are comic: it is not the preternatural as such that interests the author, but its viability as a literary device for exploring unnoticed aspects of the natural. Just as the air raids deconstruct the customary face of London, so the preternaturalist story can open up the mundane and reveal its roots in mystery.

The plot of Bowen's shortest, most lyrical, and mannered novel, A World of Love, hovers on the verge of the preternatural. A bundle of old love-letters discovered in an attic becomes the medium for the opening-up of bygone sorrows: personal relations and tensions are reactivated; a ghost, it would seem, almost materializes; and these all-too-real but below-the-surface processes are echoed in the fantasy-life of a small girl with a dream companion. Although the veil between the two worlds shivers (or, as Bowen likes to put it, 'crepitates'), it remains unbroken; and in so doing it suggests the eternal presence of the mysterium.

It is part of Bowen's essentially tragic wisdom that the alternative to the inadequacies of materialism lies not in a spiritual other-world but in hallucination. 'To be human's to be at a dead loss.'[25] While her work evinces a sharply analytic responsiveness to physical surfaces, this is matched by her brooding consciousness of human tragedy. In her world, love is foredoomed by its very compulsiveness and vulnerability. A sense of the mysterium may promise alleviation to irrationality and pain, but the imagination that apprehends it exposes its owner as much to suffering as it does to joy. Bowen's finest supernaturalist writing accordingly portrays the metaphysical as exemplary of the human, in a spirit of steadily agnostic fatalism. Ultimately her world is self-enclosed.

Phyllis Paul

One contemporary novelist whom Bowen singled out for special praise was Phyllis Paul (1903–73). Reviewing *A Cage for the Nightingale*, she notes that 'what lifts the book miles over the everyday is the language, the vision, the vehement inner poetry'.[26] Such haunting imaginative power was to impress other writers during the 1950s and early 1960s, among them John Cowper Powys, Elizabeth Jane Howard, and Rebecca West; but the times were unfriendly to the kind of novel Paul wrote, and her work failed to sustain the attention it initially received. She was not part of the metropolitan literary scene. Single, cherishing her privacy, and believing that a writer should be judged on her work alone, she paid the price such integrity exacts.[27]

Her first two novels were published in the 1930s. They have a primitive, obsessional quality comparable with that found in the contemporary novels of the Herefordshire writer, Margiad Evans, though without their violent depiction of sexuality. But it is the nine books published between 1949 and 1967 which constitute Paul's mature achievement. They differ markedly from the naturalistic, satirical novels in favour at the time. They are mysteries, not in the sense of being mystifications or enigmas (intricate and sometimes baffling though their plots can be), but in the sense of being enacted and embodied patterns of supernatural events. Despite their frequently violent content, they are not thrillers; an undertone of moral seriousness sustains them. In their sombre presentation of the absolute nature of good and evil they resemble the novels of the French Catholic, François Mauriac, though their own point of view is well-nigh Calvinistic. Paul's imagination seems to share the Protestant roots of Maturin and Le Fanu, and is imbued with a no less fanatical distrust of the workings of the Catholic hierarchy, most blatantly expressed in *Pulled Down*, with its hostile account of the movement Opus Dei. She writes out of a coherent and consistent imagination that circles round recurring themes and images, an imagination that might be castigated for morbidity. Adept at portraying derelict gardens, decaying houses, dark woods, and threatening skies, she revitalizes the Gothic novel.

The title of her first book indicates her standpoint: *We Are Spoiled* (1933). She believes in original sin. The spoiling of her characters is self-inflicted: careless upbringing, moral blindness, deliberate and envious manipulation of the weak, all serve to produce the overgrown child, the irresponsible adult, who in successive novels is to prove an agent of destruction. Moreover, those people who do preserve a spiritual chastity only attract their opposite, so that the harm done to them by the exploiters devastates the weaker characters around them. (Such an ambiguous approach to the subject of innocence recalls the work of James and Bowen: one thinks of 'The Turn of the Screw' and Bowen's *The Death of the Heart* (1938).) Again, in Paul's second novel, *The Children Triumphant* (1934), a study of spiritual crippling, the children, as in William Golding's *Lord of the Flies*, are potential savages, their vulnerable energy a menace to adult rationality and moral poise.

Camilla (1949) establishes the method, the outlook, and the themes of Paul's post-war novels. And it is post-war in its mood, not the optimistic viewpoint of the supporters of the Attlee government but the mistrustful, bleak outlook of an individualist who sees the consequences of unleashing the dogs of war, dogs that are not easily got back into the kennel. In the words of one of the protagonists, 'We're sick of blood-letting—*but how we miss it!*'

The speaker is Hartley Rupell, a former clergyman, the spoilt adopted child of a rich elderly couple. His sister Frances has been brought up by impoverished relations, and both have drifted into petty crime. Hartley exploits an easygoing philanthropic family (appropriately named Grant) by persuading them that his sister has psychic powers and has seen the spirit of Camilla, a teenage daughter of the house, who has disappeared and is believed to be dead. Frances does indeed think that she sees Camilla, but is sceptical as to hauntings or apparitions, whereas Hartley, who knows himself to have been the indirect cause of the girl's death, sees her ghost himself. The puzzle is never quite unravelled, though the author gives her readers a scattering of adroitly placed clues as to possible solutions. But Paul's real interest is in the downtrodden Frances, kindly, ironic, feckless, cherishing a gift for poetry—a gift for which, as a woman, she is despised by both her brother and the spiteful 'literary' husband who disowns her. After her death Hartley purloins a poem written as a result of her knowledge of his preoccupation with Camilla, and palms it off as his own, to critical acclaim.

Camilla is a mystery story and as such is both ominous and confusing. Paul is skilful at building up tension, exemplified in Frances's reaction to the Grants' huge London house:

she was again vaguely oppressed . . . by a thought of the many unguessed-at nooks and corners surrounding her, the closed cupboards of unknown depths, the black slits of half-open doors, the recesses, swamped in shadow, formed by projecting lumber and the disused furniture with which the place was cluttered, and the sharp, dark turn at the end of the long passage where it led down to an underground region she had never visited. (ch. 24)

But there is no climax, no sudden pounce of horror. We are in the world of the hallucinatory rather than of the preternatural. The puzzling elements are left as puzzles.

Much of the novel is told in retrospect and flashback. Paul dissolves normal chronology in favour of an imaginative rather than a temporal logic. In delving into the past lives of brother and sister one grows aware of strange coincidences, of overlappings of experience. Just as Hartley in his youth has been responsible for the suicide of a mentally disturbed older woman, so later in the novel he is responsible for the death of the emotionally unstable Camilla. At the time of the first event he is haunted by a smell of roses; at the time of the second one the roses are real. The ordinary sequences of time have in this psychic world been dissolved. Similar patternings inform other parts of the book. But this extrasensory undertow is related not to the preternatural but to the spiritual world. Paul has an inexorable vision of moral law: things are implacably as they are. *Camilla* makes a clear distinction between the spiritual and the occult. At the root of it is the awareness of the futility of easy supernaturalism and the materialistic nature of superstition. Reference is made more than once to the words 'Neither will they believe though one rose from the dead.'[28] The sombre message recurs throughout Paul's fiction.

The later novels develop themes, images, and incidents found in *Camilla*, and elaborate upon them; but for all the reiteration of symbols and plot devices, and the consistent point of view, Paul's is not a static fictive world. Although all the novels transmit a sense of the uncanny, as much through their powerful visual quality as through their depiction of superstitious fears and momentary hallucinations, only two contain specifically preternatural happenings. Of the others, *Rox Hall Illuminated* (1956) is concerned with a suppositious miracle and the growth of its attendant religious cult; *A Cage for the Nightingale* (1957) and *Pulled Down* (1964) are murder mysteries, though such a designation oversimplifies, not to say wrongs, the searching moral insight they display. *Constancy* (1951) and *A Little Treachery* (1962), clearly more personal and deeply felt, deal with mental illness and the predicament of those emotionally involved with the insane; while *An Invisible Darkness* (1967) is a mystery story first and foremost, suggestive in its outcome of Tennyson's 'Enoch Arden'. In all the novels one is aware not just of the mysterious but of the mysterium.

Of the two novels to treat specifically of the preternatural, *The Lion of Cooling Bay* (1953) is an imaginative portrait of the workings of good and evil. Less of a formal mystery story than *Camilla* or *Pulled Down*, it is relatively static: not until the apparently motiveless murder at the end is a specific action undertaken by any of the characters. The central figure is a girl called Anne, an orphan of the Blitz, who is adopted by Julian Rackenbury, eldest of a wealthy family of five brothers and sisters, renowned for their artistic interests and achievements—a portrait presumably suggested by the Sitwells. The worldly, intellectual Julian becomes obsessed with the child; and in trying to model her to his liking he treats her with what is at first unthinking cruelty. Later on his behaviour becomes sadistic, so that once she is old enough she runs away, only to fall victim to a teenage crook.

There is a dualistic subtext. Anne and the youth Francis (a characteristic duplication of the name of the protagonist of *Camilla*) are receptacles for forces of good and evil, the latter presaged by a number of preternatural manifestations. A further leitmotif is the persecution of the Albigensian heretics by the thirteenth-century Catholic Church, a subject about which Julian himself has written and with which he has become obsessed. The rich texture of the novel, its use of flashback, reverie, letters, and confessions to sustain the narrative, lends an air of genuine mystery: its people seem to be in the hands of forces far more powerful than themselves, and of which they are unaware. The moral outlook is unflinching. Anne's innocence tempts and calls out the evil latent in other people: the book presents life as a process of incessant judgement, and its message is the need for vigilance. 'Watch therefore and pray, for the devil as a roaring lion goeth about seeking whom he may devour.'[29]

For all its steely quality, *The Lion of Cooling Bay* is often humorous at the expense of the narrowly materialistic: the comedy springs from the reactions of the half-comprehending onlookers of the incipient drama. The Rackenburys are described with clear-sighted sardonic irony: their motives, so far as they consciously go, are good; but they are imaginatively crippled, being like people 'cutting up a body to find a soul'. The author is far less impressed by their artistic pretensions than by Anne's evangelistic Christianity; while she interprets natural beauty as a material cloak for spiritual reality.

A pale, an almost white sunshine had at first surrounded her, the sun's rays riding visibly on the thick air. But now the grey morning began to unfold and shine. The stretched wings of birds gleamed like skeleton fans, carrying dark bodies. Wet leaves on the privets in the little gardens glittered like silver coins, and the soot-coloured bushes were full of wet twigs which shone as if they had been split. The grass was laced and shredded and the whole scene was shot with these thin beams of light which ran along the knife-edges of stems, twigs and leaves, making a skeleton world, a world strewn with fine, white bones—a mass of tiny, bright bones was laid upon a gulf of grey air. The tangle of light falling about her seemed to have the same heightened yet undertone quality of the split, wheezing, sibilant voices of the starlings.

The language is full of violence and movement, as though the visible world were frayed and falling apart. Later the scene is to be recalled with vehement urgency, as its beholder 'suddenly remembered the road with the astonishing radiancy behind the trees at which her heart, almost shocked, had formed some thought with this sense, *"Here the light begins to burn through."*' (ch. 27). There is as much terror as beauty in the intimation.

As to the specifically preternatural happenings, rather than punctuating the narrative they interleave it. Anne's vision of a flying lion is described retrospectively in a letter; and not only that, but in the form of a summary given by Hugo Rackenbury to the assembled family after he has read it out aloud. Already her account is being misconstrued. Again, the padding of some animal in a passage in the Rackenbury house and in the streets outside, shortly before the murder, is heard by two witnesses, both of whom are puzzled but not mystified—the spiritual world impinges in vain on those who are imaginatively unresponsive. Moreover, the lion's footprints are soon discredited in the newspapers by the discovery of a hoax and the consequent facetious

scepticism. As in *Camilla*, impending spiritual presences are conditioned by the responses they elicit. This is shown effectively
through Julian's interior voice, the diabolic speaker who tells him,
as he recalls his decision to adopt a child, that 'it will not do for
you to say that you did not know her from the beginning! That,
taking this child in, you did not know whom you harboured—
and whom you thereby invited' (ch. 5). Paul's frequent use of
italics is a mark of moral emphasis, not of mere sensationalism.

In *Twice Lost* (1960) a hack writer imposes a rough-and-ready
solution to the puzzle, 'for she loved mysteries and problematical
happenings, though she did not like them to remain inexplicable,
and she loved worming things out'. But Phyllis Paul is concerned
with the mysterium. This is a tightly constructed novel, its preternatural elements still more internalized than they are in *The Lion
of Cooling Bay*; its moral stance is more compassionate, if more
exacting.

The plot works through to a firm, if open-ended, resolution. An
unattractive and neglected little girl called Vivian Lambert disappears from her home in a suburban village, and is never found:
it is presumed she has been murdered. The last person known to
have seen her alive is the 17-year-old Christine Gray, on whom
the event has devastating nervous consequences. A year later she
marries Thomas Antequin, an elderly novelist who has recently
purchased the country house in whose grounds she and Vivian
spent their last evening together. The marriage is broken up by
Antequin's son Keith, who after his father's subsequent death
finds a mutilated journal which leads him to suspect that Thomas
has been guilty of the child's death. At the same time Christine
is confronted by a young woman who claims, with some plausibility, to be Vivian herself. By the end of the book two equally
persuasive solutions to the riddle have been proposed, their clues
skilfully planted in the course of a narrative unfolded in Paul's
customary allusive manner. But the real drama is inward and
moral, and concerns self-deception. Keith is quite prepared to
blacken his father's reputation if it will further his own literary
prestige, oblivious to the fact that the aims are incompatible.
And rather than have Christine's story of Vivian's return accepted,
he would persuade her that she is mentally unstable, a conclusion to which preceding events have already made her prone. But
it is he rather than she who is deluded.

Christine is called by her husband 'a creature of light'; in this she resembles Anne in *Cooling Bay*. A gentle and affectionate woman, she is the victim of suppressed guilt over Vivian's disappearance and of a longing for certainty and peace of mind. The girl who purports to be Vivian is greeted by her as a saviour; but she in her turn disappears, leaving Christine to cry, 'Vivian, if you are dead, rise up and tell me so.' But—'neither will they believe though one rose from the dead'. When Christine realizes how her two meetings with the girl are regarded as hallucinatory her pessimism returns in force; nor can she accept her mother's confident assurance that 'God will never forsake you.' 'Tender lies. Monstrous lies. For if anything can be predicted of God, it is that He will forsake us, always, always in our worst agony' (ch. 33). Christ's cry of dereliction on the cross comes irresistibly to mind, both as confirmation and, implicitly, as reassurance; but Paul will take none of the short cuts of hopeful piety. Her next novel, *A Little Treachery*, concludes with the words, 'It happens; no one is there, no one hears.' Not even Samuel Beckett wrote anything more final. In Paul's mystery novels the wicked go free, and are not, as in detective puzzles, brought to book. In *Twice Lost* Christine's anguish may be allayed by a solution in which she feels compelled to believe, so that she 'was like a person who after long years of wandering, loaded with guilt, has at last been arrested and brought to trial, and who hardly cares that she has been miraculously acquitted, only cares that the trial is over. . .'. But the author can still conclude with the unrelenting comment, 'as she had never wanted the truth, but only comfort, so she had not now found it'. The solution of a puzzle is not the same thing as the vision of truth concealed in the mysterium.

This failure of appearance and reality to overlap extends to the novel's preternatural happenings. Even the everyday incidents have an air of strangeness: the garden of Antequin's house is portrayed with a visionary luxuriance suggestive of a Samuel Palmer painting. The preternatural phenomena are barely distinguishable from the rest; events are supernaturalized through sheer intensity of style, rather in the same way as are those in the work of Bowen. The spectral appearances relate to spiritual states. Keith's encounter with the child's ghost is the exhibition of his own character to himself. Having made cheap fun of the little girl's protruding teeth, he dies as the result of what he interprets as a bite, but really because of his own carelessness and panic. The event is both preternatural and natural, the relation between the two being the result of the victim's character. The preternatural is related to the moral law rather in the way that it is in the novels of Charles Williams; but in technique and tone *Twice Lost* looks ahead to Peter Ackroyd's *Hawksmoor*.

Paul writes out of a narrow range of experience, and when she moves beyond it she sounds as Victorian as Charlotte Brontë, the novelist to whom she bears the most resemblance; indeed, her steely puritanism suggests that here are the novels of a latter-day Lucy Snowe,[30] and her distinctive prose style, with its curiously shaped phrases, supports a similarly caustic and subversive humour. The rather shabby middle-class people she frequently describes can wear a threatening aspect, that of the semi-submerged world of the non-achievers—a world in which Walter de la Mare is similarly at home. But her understanding of the supernatural is not merely negative. 'Through all [his] pages ran a breeze of terror, arising in the metaphysical world; the air was sinister. The leaves shivered in the dark forest. Something flew behind the million leaves, always out of sight—mystery' (ch. 6). Her account of the work of Thomas Antequin is at the same time a description of what it is which makes Phyllis Paul's own writing so unforgettable and strange.

In the fiction of Bowen, James, and Paul, preternatural phenomena are internalized; they are exhibited in terms of consciousness, and are mediated through it. Terror does not impinge from without, but erupts from within. Such internalizing produces claustrophobia, perfectly exemplified in the work of James: the preternatural elements take on a paranormal quality as they are absorbed into the psyche. The psychoanalytic quest for truth, which is the motive power in fiction of this kind, seeks to lead outwards from terror, rather than inwards towards it; the thumbscrew of Gothic horror fiction becomes a mental pressure from which its victims seek deliverance in the external world. Instead of the wrath and inexplicable exactions of Nobodaddy, Blake's false father-God, and of his avenging servants, there comes the confrontation with what one has created and evoked oneself: with terror dawns the awareness of responsibility. For these writers, as for George MacDonald, the invisible world is the verdict on the world we know. Once again one sees how a supernatural

dimension in naturalistic fiction deepens that fiction's seriousness and capacity for reflecting the recurrent need of human
nature to search for some kind of absolute knowledge.

An interesting interplay is going on in these writers between
their impulsion towards harnessing preternatural material for
psychological and judgemental ends, and their own involvement
in the forces they have, as it were, conjured up. In the case of
James, the containment of those forces within the naturalistic
framework is more or less absolute. The terrors remain purely
suggestive, conveyed in almost every case at second hand: the
enclosure of one narrative within another is vital to the effect of
'The Turn of the Screw' or 'The Friends of the Friends'. But
where Bowen is concerned, the physical world itself is viewed
with such precision and intensity that, as in the case of Dickens
and, occasionally, of Hardy, it takes on properties that are little
short of numinous in their effect. The various settings of *The Heat
of the Day* seem to be steeped in the world of spirit, so that the
events, the memories, the visionary and hallucinatory experiences
that make up its people's lives become manifestations of an im-
material cosmos governed by inexorable and, where human beings
are concerned, judgemental law. This is also most especially the
case with the novels of Phyllis Paul; but in their concern with
morality and the laws of the heart, all three writers display their
affinity with those supernaturalists who work more avowedly in
a theological tradition.

8

GOD-GAMES

Under all speech that is good for anything there lies a silence that is better.

(Thomas Carlyle)

A notable aspect of the young Jane Austen's handling of belief in the preternatural is its precocious sophistication. The awareness of the diverse fictive genres displayed in her juvenilia is in *Northanger Abbey* combined with a dramatized critique of at least two others: Catherine Morland is both read about by those engrossed in this particular book, and is herself a reader of other ones, and the action of the novel constitutes a conflation of these two conditions. Austen is here anticipating late twentieth-century relationships between the author, the reader, and fictive characters in a literary mode in which the subjectivities of James and Bowen give way to an externalized and overt play of elements which would otherwise be treated as impalpable. In post-modernist fiction the dictates of naturalism are laid aside; or, rather, the constrictions are: the relativity of human apprehensions and subjective experience is formalized in a species of fictive play.

Where supernaturalist motifs are concerned, the demonic projections that form the staple of James's fiction of this kind are released into the light of day, and are severed from any tragic personal connection with human beings. They become creations of intellect, means to an end in which they play their part: their role is executive, veridical, even comic, though the comedy involved in such a presentation is closer to tragedy than it is to farce. For it is the negation of traditional supernaturalism that it should be in any way reductive of its subject.

Supernaturalists at Play

Since it will be generally acknowledged that the claim to have seen a ghost will procure immediate attention, it is incumbent upon those who make it that they should neither through facetiousness nor through avowed mendacity deprive their audience of a fright that can legitimately be passed on. Both the Montagues, Summers and Rhodes James, cite one particular type of story 'of which I disapprove'—the phrase is used by James in the prologue to his collection of Le Fanu's tales, *Madam Crowl's Ghost*, and by Summers in the introduction to *The Supernatural Omnibus*. For James, the reprehensible species was that which 'peters out into a natural explanation'; for Summers, it was 'the humorous'. R. H. Barham's 'The Spectre of Tappington', which for the latter reason Summers omitted from his anthology, will have displeased them both, since this prose opening to *The Ingoldsby Legends* disposes of its apparition with a jocularity that is at once high-handed and reductive.

The touch of a later wit proved far more delicate. In 'The Canterville Ghost',[1] Oscar Wilde brings off the feat of first breaking, then reversing, James's rule. The bulk of his story is delightful farce. Wilde does to the ghostly literary tradition what *Cold Comfort Farm*[2] does to the bucolic one: every cliché of the genre is ridiculed as the poor shade fails time and again to scare off the cheerful family who have purchased his ancestral haunt. But the fable turns serious as the burlesque gives way to the reality behind the victimized tradition. The young Virginia's disclosure of an immured skeleton, from which the ghost emanates, would neutralize the earlier amenities were not the situation then dissolved in a cloud of Wildean sentiment—that is, sentiment buoyed up by gaiety and a sense of the absurd: it even sustains the sexual overtones of the conclusion, which in hands less adroit might have seemed salacious. But 'The Canterville Ghost' entirely lacks the self-congratulatory facetiousness of its imitators.

Facetiousness is the outcome of embarrassment. Uneasy with supernaturalist ideas, the diminishing sophistication of literary taste after the Second World War resulted in credulity on the one hand and crass unbelief on the other, evident in a market for sensation novels less lurid, but more violent, than earlier varieties, and for a naturalism based as much on cynicism as on rationality. The 1950s and 1960s were a period of demythologizing and of an emphasis upon a purely humanistic attitude to life; where literature was concerned, it was an age of debunking. Accordingly, the world of Kingsley Amis's Lucky Jim and John Osborne's Jimmy Porter did not look with sympathy on the concept of the supernatural, and the literary treatment of it was largely confined to fantasies such as Tolkien's *The Lord of the Rings* and to children's books—in some estimates the same thing.

Fantasy is the supernaturalism of the agnostic, an arbitary distortion or abstraction of the normal and accepted. It does not need to contain within itself any sense of the supernatural: there is, for example, none in Mervyn Peake's invented world of Gormenghast.[3] But other novelists were to use the dislocating nature of fantasy as a means of approaching the subject of the supernatural from an oblique angle, and thus more conformably with late twentieth-century sensibilities.

Something of what these writers were to achieve is foreshadowed in *Mr Weston's Good Wine* (1927) by the Dorset novelist and short-story writer, T. F. Powys (1875–1953). The rusticity of his fictive world makes his novels and tales appear to be abstractions from ordinary life, and thus potentially allegorical in content, even when they eschew any element of the preternatural or fantastic. Powys's bucolic people and his use of dialect lay him open to parody, an invitation of which his urban critics not infrequently availed themselves; but at its best his writing has an authoritative simplicity. Powys achieves his effects less through intensity than through the quiet finality of his style, with its mannered quaintness and encompassing sardonic humour. Something of a soothsayer, he uses the homeliest imagery in order to place in an ironic light the human readiness to assume proprietorial rights over Almighty God, whom he places in his own world, embodied in a variety of figures. These include a wandering tinker, a wine merchant, a young fisherman—even, by proxy, a top hat. Side-stepping the uncanny, Powys incarnates his metaphysic in an individual person, so that the supernatural is domesticated within the allegory. In *Unclay* (1931), for example, John Death takes up residence in the village of Dodder: he is at one and the same time death personified and a man who happens (as some people do happen) to possess the surname 'Death'. One is not conscious, as one is with most allegorical fabulists, of a design upon one: it is the method rather than the content of Powys's art which provides the message.

When seen in relation to fictional events that are naturalistically described, the happenings in a fantasy or formal allegory are intrinsically artificial, overt artifice being to naturalism what the preternatural is to everyday reality. Consequently, any preternatural event within the fantasy takes on the status of the supernatural—in its teleological sense of being evidence of the laws and operations of Almighty God. Taken at a naturalistic level, *Mr Weston's Good Wine* relates the arrival of a travelling salesman in the village of Folly Down. As the result of this visit a number of misunderstandings are sorted out, couples are paired off, and two people die. The controlling preternatural event is the stopping of the clock in the village inn, a sign that time stands still. But this preternatural event is in fact a supernatural one. It constitutes the irruption of eternity into time with the arrival of the visitor, who represents God as he is operative within human understanding. The obviously fictive nature of the stylized rustic background allows for a saving irony; the supernatural operates here, as it always does, behind a mask.

It may be claimed that what Powys does with allegory is to provide a working model of how, where theological meaning is concerned, the supernatural reveals itself through the symbolism of the preternatural. His stories do not lend themselves to 'simple' allegorical interpretation; they are not dogma in disguise. Rather they are symbolic presentations of those apprehensions and operations of the mysterium which give rise to theological questions and interpretations. Such questions are naturally at home in Powys's fictive world. Within its confines the imagination can set to work at its play, as Powys in his *Fables* (1929) animates buckets, ropes, hats, hassocks, even, with supreme impertinence (or is it?) a crumb of the sacred Host. His homely animism is informed by a religious spirit. The ways of God are for man to experience, but not for man to predict or understand; as a consequence they may be detected in whatever guise one seeks to find them.

Such familiarity can appear frivolous or whimsical, to some temperaments and beliefs even blasphemous. Certainly the humour, if sedate, is mordant, as when Mr Weston remarks, in implied connection with the Bible, 'I am a writer, Mr Bird.' And it can be positively black. A good example is in one of Powys's most original and horrifying stories, 'Christ in the Cupboard';[4] the title is itself disturbing, suggestive of a Kensitite tract attacking the reservation of the sacred Host. As it turns out, the association is not misleading. On Advent Sunday a pious but greedy village family are visited by Jesus Christ, whom they recognize at once, 'for His face all the family had seen either in a church window or else in a picture'. Hoping to have Him at their beck and call, they shut Him in a cupboard; but when they let Him out they find that He has changed into the Devil. Although it has the ring of folklore, the fable is patently artificial in construction. The points are made with economy (one might call T. F. Powys a rustic Beckett). But the supernatural element does not so much lead one out into metaphysical debate as come homing in on the human application. The satirical treatment of the family's unthinking religiosity forbids any literalist interpretation: the supernatural is a symbol of the natural, not the other way round. And the cupboard? Here once again one finds the cave, the dark place, the opening into the mysterium.

Powys was writing at a time when belief in a metaphysical order that endorsed moral responsibility was on the wane; and without the tacit acceptance of such a sanction, writers increasingly turned to supernatural themes in order to write about current social or psychological concerns. Thus the inter-war years saw a proliferation of parabolic supernaturalist fables, among them being David Garnett's precisely titled *Lady into Fox* (1922) and Sylvia Townsend Warner's *Lolly Willowes* (1926), the story of a middle-aged spinster who, after years of devoted service as daughter, aunt, and sister, retires to the country and becomes a witch: in this case the uncanny takes second place to the satire and deft comic writing. The following year saw Edith Olivier's *The Love-Child* (1927), in which the dream companion of a lonely woman materializes to tragic effect. Like *Lolly Willowes* a study of the single state, the book is entirely free from the sophisticated levity of much writing of this kind. All three novels have a provincial setting, which serves to emphasize their eccentricity to the rationalistic outlook of the urban culture of their day; and a still more instructive instance of the genre is to be found in *Miss Hargreaves* (1940) by Frank Baker (like Sylvia Townsend Warner a member of Arthur Machen's circle).[5] In this story, two young men from a cathedral town, acting upon a gamesome impulse, invent an elderly spinster, whom they christen Constance Hargreaves. Pushing the fantasy to extremes, one of them writes to her, inviting her for a visit. She materializes. The rest of the book tells of his struggle to rid himself of an encumbrance he comes increasingly to love. The author steers a skilful course through the shoals of implausibility: other people encounter the old lady, who establishes herself in cathedral social circles and even buys a house. The fact that she is an invention is given a disquieting twist when she is disowned—'You can do what you like from now on!'—and she becomes an overtly hostile force, a monster capable of rending her inventor: a gentle whimsy has swelled into a latter-day version of *Frankenstein*.

Both theological and scientific explanations are proffered for the materialization; but as the local priest declares, 'Supernatural phenomena cannot be proved by natural evidence.' *Miss Hargreaves* is at one level a parable of creativity. Whatever her inventor decrees that his creature shall do, occurs. Yet Constance Hargreaves being a person with a character of her own, his powers are limited; and when she takes up the autonomy he grants her, he suffers for it. It is the relationship between the two, delicate, uncomprehending, often painful, which gives the book its special flavour. 'Who was the haunter, who the haunted?' The final disposal of the old lady is the resolution of that dilemma, and raises the protagonist's experiences to the category of the supernatural with the acceptance of responsibility and the readiness to complete the fictive drama by providing it with meaning.

Such fantasy novels enjoyed considerable popular favour in their day and exhibit a preoccupation with the potential power of imaginative creation, a half-willing acceptance that the world of logical, rational deduction may not provide the only means of controlling and ordering the material processes of the world; but they come across as particular instances and are presented with a dusting of protective irony. The altogether bolder attitude to such uses of supernaturalist fantasy that emerged during the post-war years is exemplified in *The Magus* (1966) by John Fowles (b. 1926). Although not dealing with the supernatural as such, the book does raise speculative questions as to the nature of fictive reality, through the mechanisms of the mystery novel, the occultist novel, and the tale of terror. The author at one time thought of calling it 'The God-Game'.[6] Its narrator, Nicholas Urfe, a disillusioned womanizer, takes up the post of schoolmaster on a small Greek island, and there falls under the spell (at one point literally, through hypnosis) of a wealthy eccentric called Conchis, who engages him in a series of manipulative games, providing scenarios for him to act out, challenging his sense of objective reality. Conchis acts the role of novelist, controlling the destinies of his characters, and traps young Nick into becoming a fictional character himself.[7] The latter is subjected to a series of moral and emotional shocks, which include a number of what at first sight seem to be preternatural experiences. Finally he undergoes a mock trial and is psychoanalytically arraigned; then, having suffered the depths of personal outrage, he is freed to judge his accusers in his turn. This assignment he is unable to carry out, since he finds himself irrevocably a part of what they are and have been doing.

Conchis provides an artificially induced enigma which, while it can be decoded like a puzzle, can only be experienced as mystery. The situations in which Nick has to take part are only relatively true—they constitute what Conchis calls mystery-at-noon, and take place on the assumption that 'in our century we are too inured by science fiction and too sure of science reality ever to be terrified of the supernatural again'. The terrors the story provides are terrors of the mind—mistrust of the senses, of one's power of moral judgement, of the value of making moral assessments at all. Through stressing the relativity of all these contrivances in a kind of enacted history of Western consciousness, Fowles directs attention to the transcendence of the mysterium.

The book has tremendous narrative propulsion, being written with a command of naturalistic description and a sure sense of dramatic timing. What remains in question is the exact moral status of Conchis's elaborate god-game. Nick himself is in need of spiritual rebirth, not least on account of his treatment of the girl with whom it would seem to be the purpose of the god-players to reunite him. There is something repellent in the methods they employ. In the end they withdraw, leaving the lovers free to work out their own salvation; but there is little of the angelic or benevolent about them. One's sympathies are inevitably (this being a first-person narrative) with Nick. There is a smugness about Conchis's cat-and-mouse procedures: he is not a trustworthy moral arbiter, since the book nowhere makes it clear whether he is at any time acting in good faith or not.

None the less, it is he who voices what is the novel's most valuable metaphysical insight. During his (alleged) involvement with the torture and execution of some Greek resistance fighters, he encounters in one of them something which, he tells Nick, is for him an absolute that justifies the term.

He spoke out of a world the very opposite of mine. In mine life had no price. It was so valuable that it was literally priceless. In his, only one thing had that quality of pricelessness. It was *eleutheria*: freedom. He was the immalleable, the essence, the beyond reason, beyond logic, beyond civilization, beyond history. He was not God, because there is no God we can know. But he was the proof that there is a God that we can never know. He was the final right to deny . . . he was something that passed beyond morality but sprang out of the very essence of things—that comprehended all, the freedom to do all, and stood against only one thing— the prohibition not to do all. (ch. 53)

We are back in the world of *Melmoth the Wanderer*, a novel to which *The Magus* bears some resemblance. Fowles enunciates mid-century Western civilization's stress on the absolute value of a freedom that is bound up with the very nature of Being itself. It is here that people are God-like: God-*like*, however, not God. The rejection of false and limiting concepts of the supernatural leaves the soul free to apprehend its reality at the level on which the supernatural operates. The free person encounters, and moves within, the mysterium.

Fowles's novel is an instance of that concern with perception, self-delusion, and the elusive nature of objective reality which has prompted a number of English novelists to make their own incursions into, and adaptations of, the territory of supernaturalist fiction; but whereas Fowles makes use of the paranormal to enhance the dramatic effect of his novel, the work of William Golding, Muriel Spark, and Iris Murdoch deploys traditional preternaturalist imagery to ends at once more searching than his, and more concerned with a positive approach to disillusionment. For if disillusionment means the loss of illusions, it is not an experience to regret.[8]

William Golding

The recent increase in professional critical appraisals of imaginative writers by their contemporaries has contributed in part to a corresponding increase in the reputations of those novelists whose work lends itself most readily to such attentive scrutiny. Academic critics welcome an author whose work they can expound, even more than they welcome one whose work they can deconstruct.

In this respect both John Fowles and William Golding (1911–93) have had their full share of expository attention, and in both cases their novels have themselves encouraged it, being avowedly concerned with the craft and epistemological standing of fiction, and with experimentation that may well owe as much to the critic on their heels as it does to any quest for an absolute form of fictive statement. But Golding's work also circles round certain metaphysical preoccupations.

The early adoption of *Lord of the Flies* as a school set text, the proliferation of commentaries and studies, witness to his being a novelist whose work demands both to be decoded as a puzzle and pondered as a mystery. And yet none of his books contains any elements of the preternatural: powerful, even eerie, though certain scenes, encounters, and experiences are, what they provide are instances of the paranormal that, in the context of each novel as a whole, can be read as aspects of the supernatural. Few, if any, English novelists employ so multi-layered or complex a methodology.

The early novels come in pairs, and complement each other. *Lord of the Flies* (1954) portrays the loss of innocence, through the image of civilized man becoming a savage; *The Inheritors* (1955) through that of prehistoric man becoming civilized. *Pincher Martin* (1956) describes one man's fallacious attempt at all costs to keep his freedom; *Free Fall* (1959) another man's attempt to discover how he lost it. *The Spire* (1964) describes the birth of aspiration, and *The Pyramid* (1967) its death; the former is dominated by the image of life arising out of a death founded in sin and pride; the latter is dominated by the image of a tomb, the death of innocence and trust. But such interpretative couplings, however detectable, are misleading; each novel subverts the attempt towards a definitive reading. It is as though the plot and the plot's significance are deliberately set at odds, and yet in *Lord of the Flies* the reversal of perspective on the last page, whereby the 'savage' boys are seen through adult eyes, does not really refute the imaginative realities conjured up by the rest of the book. And the whole point of *Free Fall* would seem to lie in the fact that Sammy Mountjoy's quest for a precise moment to blame for the loss of his freedom is rebutted by the Commandant's words, which close the novel, 'The Herr Doktor does not know about peoples.' Human judgements are always and necessarily relative. And while these early novels take a specific literary text or myth as model, they are also a refutation of the sufficiency of books. The theme is continued in the later works, in the mutual self-destruction of author and exegete in *The Paper Men* (1984) and the continuous learning process charted in the *To the Ends of the Earth* trilogy (1980–9). In every case, imagination and delusion are in constant interplay.

Where the supernatural is concerned, two texts discuss the matter explicitly—*Lord of the Flies* and *The Spire*. In the former, Simon's dialogue with the diabolic image erected by the hunting party amounts to a refusal of the preternatural through an assertion of material reality: he rebuts the temptation to give way to fear by the plain statement, 'Pig's head on a stick'. This complements the boys' notion of 'the Beast' when the dead pilot and his parachute are discovered on the mountain and their fear of the preternatural leads to fear of the self.

In *The Spire*, Jocelin's 'angel' is in fact the manifestation of spinal cancer: his belief in his vocation to build the cathedral spire would therefore seem to be discredited. But on a supernaturalist reading of his experience, this is not necessarily the case. The spire is founded in both good and evil motives, in both paganism and Christianity, in courage and obstinacy, disinterested vision and arrogant self-serving; what matters is its existence as a thing in itself. 'It's like the apple tree', is what Jocelin comes to realize: what matters is the living world, not his own particular part in it. Golding's controlling outlook, bleak though it is, is comprehensively religious. And the intricacy of his novels' organization, their complex imagery, multi-layered narratives, and embodied paradoxes, all underwrite this sense of the supernatural, even while the propulsion of each book discourages the location of that sense in any one particular element within the text.

This perspective is itself subject to a gloss in Golding's most despairing novel, *The Paper Men*. Here the cynical narrator Wilfred Barclay appears to receive the stigmata, only to be reminded by the local parish priest that not only Jesus was crucified, but two thieves with him: his scepticism as to the miracle is thus ironically outflanked. And the concept of the eternal present, ordinarily so consoling, is here subjected to scrutiny as being itself the product of time, choice, the process of human history, so that 'link by link, we don't know what will come from this seed, what ghastly foliage and flowers, yet come it does, presenting us with more and more seeds, millions, until the whole of *now*, the universal Now, is nothing but irremediable result' (ch. 5). The sense of judgement is implacable: in this, one of his more ostensibly amusing books, Golding's vision is uncompromisingly dark.

More ambitious and less disheartening is his most perplexing novel, *Darkness Visible* (1979). Here the puzzle elements of the earlier novels give way to a more confident exposition of mystery. At one level the book is as obviously fictive as were its predecessors in the naturalistic mode, *Free Fall* and *The Pyramid*. In all three the use of blatantly allegorical names for places and people (Stillborn, Greenfield, Father Watts-Watt, Beatrice Ifor, and so on) alert one to their not being 'a slice of life' but a proffered model of life.

Even on the surface level, *Darkness Visible* is cryptic, with the mysterious Matty, a horrifically scarred religious visionary, thwarting a terrorist plot in the moment of his death, and, after that death, bringing release to the compulsive pederast, Sebastian Pedigree: indeed, were the book to be judged at the level of plot alone, it would be so inchoate as to be meaningless. But when taken allegorically the pieces start to come together: the humanists are named Goodchild and Bell, the beautiful twins, those false innocents, are called Sophia and Antonia, types of the Eastern and Western Roman Empires, of wisdom and rule, however perverted. Matty's name is Matthew (author of the Gospel of the New Law) Septimus (seven, the mystic number of completeness) Windrove ('the wind bloweth where it listeth'—an image in the Fourth Gospel (John 3: 8) for the operations of the Holy Spirit). Sebastian Pedigree likewise is aptly named—the boy victim and favoured saint of homoerotic artists, coupled with the sense of ancestry, chronology, time, and predestination. Matty, who compulsively enacts expressive rituals similar to those performed by the Old Testament prophets, is a type of the Suffering Servant, a man 'without comeliness and acquainted with grief' (Isaiah 53: 3); he is a redemptive Christ figure, whose origins recall the poem 'The Burning Babe' by the sixteenth-century Jesuit, Robert Southwell. In Australia, down under in a symbolic Hell, he is grotesquely crucified by an aboriginal, type of the old Adam or fallen man. And so on: the possibilities for exegetical ingenuity seem endless. Matty himself, however, is not an allegorist, but a thoroughgoing literalist. Golding gives his journal entries a directness and simplicity suggestive of Bunyan or George Fox, so that one is persuaded both of the reality of his visions and of the essential truthfulness of what he knows and proclaims. We are also shown the limitations of that vision, in Matty's feelings of guilt where Pedigree is concerned, a guilt that is absolute for him but relative to the 'facts' of the case as presented in the account of their relations. The author incorporates a genuine supernatural experience into a realistic narrative without any sense of contradiction.

At the root of Matty's vision is the need for silence, a silence that will offset the noise of twentieth-century civilization, imaged here not only in the juggernauts that thunder through the town and the jet aircraft that swoop above it, but also in the secondhand volumes in Sim Goodchild's bookshop, 'full of words, physical reduplication of that endless cackle of men'. However, the theophany that Sim experiences in the seance conducted by Matty reveals itself in sound.

It was a single note, golden, radiant, like no singer that ever was. There was, surely, no mere human breath that could sustain the note that spread as Sim's palm had spread before him, widened, became, or was, precious range after range beyond experience, turning itself into pain and beyond pain, taking pain and pleasure and destroying them, being, becoming. It stopped for a while with promise of what was to come. It began, continued, ceased. It had been a word. That beginning, that change of state explosive and vital had been a consonant, and the realm of gold that grew from it a vowel lasting for an aeon; and the semi-vowel of the close was not an end since there was, there could be no end but only a re-adjustment so that the world of spirit could hide itself again, slowly, slowly fading from sight, reluctant as a lover to go and with the ineffable promise that it would love always and if asked would always come again. (ch. 13)

It is a sardonic reminder of the trivializing power of the contemporary world that this moment, having been accidentally televised by the police should become a source of mockery to the viewing public. The place where the seance is held, with its message that would overset the ordinary values of the world, is coincidentally the place where a crime is plotted that would overset ordinary human values in quite another way.

Golding's portrayal of the emergence of evil in the consciousness and life of the girl Sophie is explicit and convincing: her response to the notion of entropy, her collusive fatalism, are a fitting twentieth-century version of the kind of romantic disappointment voiced in a novel like *Melmoth the Wanderer*. But the heartlessness is worse, for there is no metaphysic to underwrite it. As Pedigree says,

they did it while they were young. Willing to kidnap a child—not worrying who got killed—imagine it, those young men, that beautiful girl with all her life before her! No, I'm nowhere near the worst, gentlemen, among the bombings and kidnappings and hijackings all for the highest of motives . . . (ch. 16)

The darkness is all too visible in this novel, yet the vision is not despairing; for all the dark there still is light. Golding, a true supernaturalist, is able to envisage a totality underlying the fragmented nature of human experience. As in Matty's vision in the bookshop window, we are shown 'the seamy side where the connections are. The whole cloth of what had seemed separate now appeared as the warp and woof from which events and people get their being' (ch. 3). The use of the true meaning of the word 'seamy' is a piece of genuine, transforming illumination: this radiant awareness of the essential order of the cosmos illuminates, however fitfully, all Golding's novels. But he refuses any simplistic, acquiescent quietism. For him the Eternal Now is, as often as not, an occasion for anguish, as it is *The Paper Men;* and in the same novel Barclay's vision of Christ is scarcely reassuring.

It was a solid silver statue of Christ but somehow the silver looked like steel, had that frightening suggestion of blue. It was taller than I am, broad-shouldered and striding forward like an archaic Greek statue. It was crowned and its eyes were rubies or garnets or carbuncles or plain red glass that flared like the heat in my chest. Perhaps it was Christ. Perhaps they had inherited it in these parts and just changed the name and it was Pluto, the god of the Underworld, Hades, striding forward. I stood there with my mouth open and the flesh crawling over my body. I knew in one destroying instant that all my adult life I had believed in God and this knowledge was a vision of God. Fright entered the very marrow of my bones. Surrounded, swamped, confounded, all but destroyed, adrift in the universal intolerance, mouth open, screaming, bepissed and beshitten, I knew my maker and I fell down. (ch. 11)

The collocation of this distorted vision with Barclay's attack of sunstroke (appropriate enough in any case) is exact. The red eyes with their demonic associations preserve a certain ambiguity. But that 'intolerance' is Golding's sense of things as they are. It is one of the late twentieth century's most impressive witnesses to the true meaning of the supernatural.

Iris Murdoch

Had he never written them, one almost feels that twentieth-century critics would have had to invent Golding's novels, so readily do they lend themselves to exegesis. The more general popularity of those of Iris Murdoch (b. 1920), on the other hand, is a tribute to her command of straightforward storytelling. For all their philosophical content and their rigorous morality, they are filled with the Gothic settings and motifs beloved of the readers of fiction from the time of Ann Radcliffe to the present day. The secluded abbey by the lake, the huge isolated fog-bound rectory, the faceless terraces of outer London, might have appeared in the novels of Collins, Le Fanu, or Machen; yet Murdoch describes these places with a wealth of naturalistic detail that makes them indisputably her own creation. Like those of Henry James or Elizabeth Bowen, her characters are drawn from a fairly restricted social group, but they recur in a variety of combinations as part of a particular microcosm, and have been well described as 'deceiving husband, complacent middle-aged wife, troubled late-adolescent, middle European Jew, *homme moyen sensual*, refugee, arty mistress, scholar, would-be waif, Peter Pan boy-woman, honourable soldier, glossy civil servant, witch, demonic girl-child, outsider, secret homosexual, failed writer, dabbler in eastern religions'.[9] The list illustrates the blend of naturalistic and mythological material which distinguishes her imaginative world.

Murdoch adapts and extends the fantastic tales of Stevenson and Chesterton. Her characters are given to wild extravagances of behaviour and her stories to bizarre twists and confrontations as a woman's clothing is systematically slashed to ribbons, a fox leaps into a parked car, a man is raised from the dead, people are trapped in caves, swing on chandeliers, practise black arts in office basements, make suicide pacts, change their sexual partners and their sexual habits, get drunk, enjoy religious ecstasies, wrestle in coal cellars, attend seances, and unendingly deceive each other. Murdoch's world is opulent, generous, and emotionally exacting. She may repeat her effects, but she is forever ringing the changes on old themes. Her novels are as full of puzzles, coincidences and cliff-hanging chapter endings as are those of Hardy; while her sense of the hyper-natural and the rich texture of her surface narratives challenge comparison with the fiction of

Dickens and John Cowper Powys. Like theirs, her novels contain frequent elements of the preternatural.

These range from magical rituals to random and inexplicable phenomena, such as the mysterious marine creature in *The Sea, The Sea* (1978) or the silent gypsy who keeps appearing and disappearing in *The Sandcastle* (1957): in neither case is an explanation forthcoming. The one may be a delusion, the other a coincidence: Murdoch's world is sometimes as uneasy and ambiguous as that of Phyllis Paul. But it has a far wider social ambience, and embraces such a variety of sensuous and aesthetic pleasures and responses that these preternatural happenings are gathered up into a total view of phenomenal existence that is both rich and strange.

What makes Murdoch's writing distinctive is the combination of exceptional readability with the intellectual subtlety exerted by James or Bowen at their most exacting. Some of the novels, such as *The Black Prince* (1973) and *Henry and Cato* (1976), make uncomfortable demands: the depiction of moral and metaphysical imperatives is unflinching. At the end of the latter book Brendan Craddock says of the quest for absolute truth,

> one will never get to the end of it, never get to the bottom of it, never, never, never. And that never, never, never is what you must take for your hope and your shield and your most glorious promise. Everything that we concoct about God is an illusion.

The apparently bleak message is couched in formal, slightly archaic language: the author's sensitivity to contemporary thought and feeling does not prevent her from relating them to earlier, equally valid responses.

One of her finest portrayals of a loss of religious faith and of its resolution comes in *Nuns and Soldiers* (1980). Anne Cavidge has left the convent to which she has withdrawn in 'a flight of innocence'. Her exposure to the lacerations of unrequited passion purges and refines, without ultimately lessening, her capacity for belief in the sovereignty of love: in portraying her Murdoch reveals a remarkable understanding of religious psychology. Especially effective, and a good instance of her tactful intelligence in handling supernaturalist material, is the scene in which Anne has a visionary encounter with Jesus Christ. Beginning as a dream, it changes to 'a veridical vision' which leaves behind it as tangible witness an elliptical chipped stone and a scar on Anne's finger where she has touched the visitant's shirt. The dialogue is made up of biblical echoes and of overlappings with the words of Julian of Norwich, in itself the product of Anne's mind and reading, and yet true to an objective world that is the correlative of her own imagination. The encounter, left open as to its ontological status, returns Anne to the responsibility of adult belief, to a faith that asks for no props or proofs. 'Love me if you must, my dear,' the Jesus figure tells her, 'but don't touch me.'[10] The phrase encapsulates both Murdoch's charitable vision and the rigour of her philosophical understanding of the supernatural.

Her literary presentation of this moral point of view is bound up with her awareness of the perils of magical delusion. *The Unicorn* (1963), one of her more paradoxical and perplexing works, is centred on this theme. The setting is an isolated house on a wild and lonely sea-coast. A young woman with the decorous Victorian-sounding name of Marion Taylor, arrives to act as companion-cum-governess to a wealthy recluse called Hannah Crean-Smith, who seems to be held a not unwilling captive by the absent husband whom in the past she has betrayed and tried to murder—or so the story runs. Together with Hannah's London admirer, Effingham Cooper, Marion tries to rescue her, with catastrophic results: at the end of the book both Hannah and her principal guardian are dead, while the two outsiders return sheepishly to the metropolis, having left a trail of destruction behind them. The conclusion resembles that of Forster's *Where Angels Fear to Tread* (1905), in which officious good intentions are equally self-cancelling.

The succession of coincidences and reversals is inconsequent and bewildering enough to suggest that the author has something in mind other than neat foreclosure. For if *The Unicorn* is read as a straightforward symbolic narrative, with a master key to unlock the various illustrative devices, it emerges as a muddle, undisciplined and structurally incompetent. While seeming to ask for an allegorical reading, it resists one. The riddle of Hannah's true nature is left unsolved: we can never be sure whether she is an evil enchantress or an innocent victim, a scapegoat or a killer. All the characters live with interpretations of her which in turn reflect their own deeds and natures; but not one of them undergoes a permanent transformation. So where is reality? Is it in the spiritual drama, ritually enacted, which is going on in the lonely house? Or is it in the outer world from which Effingham and

Marion are visitants? As Murdoch presents the matter, the two modes of discourse are not comparable; but neither is the mythic world of the unicorn to be seen as a judgement on the world of affairs. Since it is itself confused, destructive, and open to abuse, it cannot provide an infallible scripture according to which the operations of the outer world are to be read. The novel is itself a commentary on other novels of the kind. It suggests that all patterns, mythical, psychological, ideological, are relative: we are all at the mercy of our subjectivities.

This is made apparent in the book's single most arresting moment. Effingham wanders out into the boggy wilderness that lies beyond the house, and becomes lost there and benighted; he is rescued from drowning in a swamp by a native of the district, who appears to be walking upon the water like Christ, as he brings a donkey (another Christ-figure association) along the one safe path to carry the lost wanderer home. Effingham's sense of gradual entrapment is grippingly conveyed—Murdoch is adept at the kind of descriptive writing whereby an experience is heightened to an intensity of concentration that seems to take the subject of it out of time. Still more memorable is the epiphany that attends Effingham's approach to death. He asks himself whether, when he is dead, anything will be left.

It came to him with the simplicity of a simple sum. What was left was everything else, all that was not himself, that object which he had never before seen and upon which he now gazed with the passion of a lover. And indeed he could always have known this for the fact of death stretches the length of life. Since he was mortal he was nothing and since he was nothing all that was not himself was filled to the brim with being and it was from this that the light streamed. This then was love, to look and look until one exists no more, *this* was the love that was the same as death. He looked and knew with a clarity which was one with the increasing light, that with the death of the self the world becomes quite automatically the object of a perfect love. (ch. 20)

Yet even this epiphany is not in the last resort determinative: Effingham forgets it, or rather, cheapens it, in the flurry of his cosseting by the women of the house on his return. Nevertheless, it exemplifies the best that most of the aspirant characters in Murdoch's novels can rise to. The natural world, properly perceived, is supernatural. It is the quality of the human response which determines whether or not it is recognized as such.

If *The Unicorn* has a message it is an ambiguous one. On the one hand it asserts the power of the imagination to project myths and fantasies in order to bring meaning to human existence; without them we are not human at all. At the same time these symbols and conceptions are the product of our natures, and cannot be treated as definitive or absolute. What post-structuralist critics posit of critical absolutes, Murdoch posits of a novelist's material. In *The Unicorn* she portrays a group of people who try to impose authoritative interpretations of their own making upon the uncontrollable processes of chance events. They seek to force and to restrict the allegory of life, and are thus imaginatively literal-minded. Throughout her fiction it is Murdoch's compassionate indictment of many of her characters that, in their quest for reality, they insist on playing games of their own devising, and thus narrow life's possibilities to essentially destructive ends. The moral drive of all her novels is thus towards purging the mind of cant, and towards cleansing the imagination from an idolatrous worship of its own ideals and images: the various failures and humiliations undergone by her people are all to do with their confusing the various relativities of this world with an absolute attested by exposure to the mysterium, and yet at no time and in no manner to be placed or named. The most they can hope for is to repent in the face of what can at the best be designated 'a fairly honourable defeat'.

The book with that particular title marks the beginning of Murdoch's most impressive period of creativity. The pattern of relativities is now stretched over a leisurely amplitude of characters and incident: the overtight patterning and structures of the earlier novels give way to the recognition that patterns are themselves relativities which, when imposed as absolutes, even in the shape of literary form, can be denaturing. The fact that in one of the most Gothic of the earlier books, *The Time of the Angels* (1966), Murdoch should have envisaged the possible demonic consequences of such a refusal of dogmatic certitude only goes to make her embrace of such polyphonic novelistic structuring the more impressive.

Perhaps the key novel for understanding Murdoch's fictional purposes is the one that followed, *The Nice and the Good* (1968). In it the distinction between those two concepts is clearly drawn, with a dramatically effective contrast between an idyllic country house on the Dorset coast, peopled by those who live under the tolerant and somewhat complacent care of its happily married

owners, and a seedy world of occultism and blackmail located in
the capital: its opening recalls Charles Williams's *War in Heaven*.
This book contains one of the author's specific descriptions of
the paranormal.

There are mysterious agencies of the human mind which, like roving
gases, travel the world, causing pain and mutilation, without their owners having any full awareness, or even any awareness at all, of the strength
and the whereabouts of these exhalations. Possibly a saint might be
known by the utter absence of such gaseous tentacles, but the ordinary
person is naturally endowed with them, just as he is endowed with the
ghostly powers of appearing in other people's dreams. So it is that we
can be terrors to each other, and people in lonely rooms suffer humiliation and even damage because of others in whose consciousness perhaps they scarcely figure at all. (ch. 17)

This is reminiscent of *A Glastonbury Romance*. But Murdoch
gives the insight a more sinister and tragic twist than Powys
does: it is out of such a consciousness that some of her most
eerie and disturbing effects arise—the spirit-haunted mind of Tallis
Browne in *A Fairly Honourable Defeat* (1970), for example, or the
hold of Beautiful Joe over his priestly victim in *Henry and Cato*,
or the poltergeist nature of Austin Gibson-Grey in *An Accidental
Man* (1971), or James Arrowby, the Buddhist adept, in *The Sea,
The Sea*.

This last is the finest of Murdoch's supernaturalist novels because the most comprehensive. While some of the narrator Charles
Arrowby's preternatural experiences in that most convincing of
haunted houses, Shruff End, are either materially accounted for
or are explicable in paranormal or magical terms, the relation
between his psychological and moral condition and his vision of
the sea-monster (let alone his glimpse of a woman's staring face
in an empty room at night) suggests a reciprocity between spiritual and physical experience which is of the essence of the
supernatural. Moreover, his cousin's occult powers (which James
himself refers to as 'tricks') are accepted for what they are, and
thus quite naturally become part of the total fabric of the novel,
as much subject to moral and intellectual scrutiny as are the
more formally naturalistic elements. *The Sea, The Sea*, with its
weird landscape and its searing analysis of obsessive jealousy,
conveys the authentic shiver of recognition at the presence of
mystery, whether it be in the uneasy psychic vibrations at Shruff
End, or in Charles's blissful vision of celestial harmony on the

cliffs at night. The supernatural, in both its destructive and its affirmative aspects, governs the action of the novel.

From this point of view, James Arrowby's magical powers are more dangerous than beneficial, despite his use of them to rescue Charles from drowning—a good instance of Murdoch's adept hand in describing what would normally be regarded as an unbelievable occurrence. James himself is aware of the temptation to worship power even as he succumbs to it; and it is he who enunciates a truth which the novel itself exemplifies: 'White magic is black magic . . . Demons used for good can hang around and make mischief afterwards. The last achievement is the absolute surrender of magic itself, the end of what you call superstition.'

On balance it is Murdoch's vision of supernatural good that is impressive, rather than that of supernatural evil. She is no metaphysical dualist. John Ducane's reflections on the dabblers in black magic in *The Nice and The Good* indicates her moral and intellectual priorities: 'Perhaps there were spirits, perhaps there were evil spirits, but they were little things. The great evil, the dreadful evil, that which made war and slavery and all man's inhumanity to man lay in the cool self-justifying ruthless selfishness of quite ordinary people . . .'. However, Murdoch's sense of the supernatural dimension is more positive.

Each human being swims within a sea of faint suggestive imagery. It is this web of pressures, currents and suggestions, something often so much less definite than pictures, which ties our fugitive present to our past and future, composing the globe of consciousness. We think with our body, with its yearnings and its shrinkings and its ghostly walkings. (ch. 39)

This might be described as the paranormal dimension in which the good is operative: but the novel closes with a more transcendent epiphany, the twin children's vision of one of the flying saucers which only they can see.

The shallow metal dome glowed with a light which seemed to emanate from itself and owe nothing to the sun, and about the slim tapering outer extremity a thin line of lambent blue flame rippled and leapt. It was difficult to discern the size of the saucer, which seemed to inhabit a space of its own, as if it were inverted or pocketed in a dimension to which it did not quite belong. In some way it defeated the attempt of the human eye to estimate and measure. It hovered in its own element, in its own silence, indubitably physical, indubitably present and yet other.

A contemporary piece of popular mythology is fused with an ageless concept of the metaphysical, without portentousness and even with a touch of intellectual gaiety.

Muriel Spark

A flying saucer of a different kind is to be found in a tale by Muriel Spark (b. 1918), when a small one invades the home of an antique-dealer. The idea is a typical English whimsy—Sylvia Townsend Warner might have thought of it—but the resultant argument is characteristic of Spark's peculiar astringency. A friend thinks the saucer might be radioactive.

'It is not radio-active,' said Miss Pinkerton, 'it is Spode.'[11]

The disagreement escalates; the saucer vanishes. The effect is dislocating in a double sense.

On the surface Spark's work is full of oddities, strange in the picturesque way that Murdoch's incidents are often strange; but Spark writes a leaner prose and has a more sinewy imagination. On a first reading her novels seem slick and heartless; she makes no concessions to nostalgia or sentimental optimism, or to those people who 'always look at what might be, or what should be, never at what is'.[12] The nature of 'what is' forms Spark's main preoccupation. A convert to Catholicism, she takes the metaphysical dimension for granted; the eternal now is present in her fiction by implication, and by her steady subversion of materialistic assumptions. The economy and austerity of her literary methods should not blind one to the complexity of their results.

Spark has always been concerned with the nature and role of fiction as a corrective to the limitations of materialistic concepts of the truth. At the end of one of her later and more amiable books, *Loitering with Intent* (1981), the narrator, herself a novelist, remarks that 'when people say that nothing happens in their lives I believe them. But you must understand that everything happens to an artist; time is always redeemed, nothing is lost and wonders never cease.' People in Spark's fictive world exist by their fantasies, obsessions, imaginations, compulsions, much as they do in the novels of Dickens, Iris Murdoch, and John Cowper Powys; but Spark is more concerned with the metaphysical implications of the human phenomenon. Her attention is fixed on whether it is radioactive rather than whether it is Spode.

Her first novel, *The Comforters* (1957) combines an inquiry into the nature of fictive truth with an inquiry into the nature of supernatural reality. That the book should also be a comedy is indicative of her particular skills: she is the airiest of fabulists, the least portentous of supernaturalist writers. In this novel can be found the elements and characteristics of her later ones, while at the same time it adumbrates their distinctive point of view.

Caroline Rose, at work on a book about Realism in the English novel, is haunted by the sound of a typewriter and a woman's voice reading out both her own thoughts and a narrative of the events in which she is concurrently taking part. Coming to the conclusion that she herself is the subject of somebody else's novel, she is forced to question her own reality, even attempting to prove her own free will by thwarting the unseen author's plans for her—an aim in which she fails. At the end of the book she is herself writing a novel, which presumably is the one we have just been reading—a Proustian hat trick. *The Comforters* is a neat parable about Divine Providence, and about human response and responsibility.

Caroline found the true facts everywhere beclouded. She was aware that the book in which she was involved was still in progress. Now when she speculated on the story, she did so privately, noting the facts as they accumulated. By now, she possessed a large number of notes, transcribed from the voices, and these she studied carefully. Her sense of being written into the novel was painful. Of her constant influence on its course she remained unaware and now she was impatient for the story to come to an end, knowing that the narrative could never become coherent to her until she was at last outside it, and at the same time consummately inside it. (ch. 8)

The parallel plot describing the activities of a gang of smugglers in a seaside village introduces a puzzle element in contrast with the mystery of Caroline's apparent hallucinations. The novel also contains an element of preternatural fantasy, as when the self-righteous Mrs Hogg disappears out of the back seat of a moving car as soon as its two other occupants forget about her. As one of them says of her more than once, she is 'not all there', and Caroline, thinking of the nature of fiction, decides that she is 'Not a real life character . . . only a gargoyle.'[13] The author herself (Caroline?) comments, 'She had no private life whatsoever. God knows where she went in her privacy.' The colloquialism is theologically exact.

Spark plays these games with great skill. At one point Caroline in hospital informs an old pseudo-diabolist that

The Typing Ghost has not recorded any lively details about this hospital ward. The reason is that the author doesn't know how to describe a hospital ward. This interlude in my life is not part of the book in consequence. (ch. 8)

The author then goes on to comment that 'It was by making exasperating remarks like this that Caroline Rose continued to interfere with the book.' And then on the next page one reads just such a description, meticulous and substantial: is this Caroline's work or the author's?

Everyone in the novel holds beliefs which to the others are 'fantastic'. Caroline herself defines the distinction between the preternatural and the supernatural in terms of religious faith and self-authenticating personal experience.

You are asking me to entertain impossible beliefs: what you claim may be true or not; I have doubts, I can't give assent to them. For my own experiences, however, I don't demand anyone's belief. You may call them delusions for all I care. I have merely registered my findings. (ch. 8)

The Comforters makes its points through the ambiguities of comedy, wordplay, a juggling with past and present tenses, a deliberate artificiality; to this extent it sets the tone for most of the author's later fiction. In *Memento Mori* (1959) and *The Ballad of Peckham Rye* (1960) abnormality again intrudes, but the humour is less playful. *Memento Mori* describes the reactions of a number of elderly people to random telephone calls reminding them that they must die. The calls are treated by most of them, and by the police, as a puzzle; but they are a mystery, each person hearing a different kind of voice, the multifarious voice of, by implication, God. The theme is treated quite without portentousness or solemnity: the book frequently borders upon farce. Also upon bad taste: its depiction of the physical indignities of age is chillingly exact. Indeed, one feels inclined to say that *Memento Mori* is a novel which no one over 60 should read and that no one under 60 should have written.

In *The Ballad of Peckham Rye* the Devil—or a devil—is abroad in the person of Dougal Douglas (or Douglas Dougal), a young man with what appear to be amputated horns (nothing is absolutely certain in Spark's fantasies) and with a horror of unhappiness and sickness: he makes mischief like some suburban version of Evelyn Waugh's Basil Seal. This becomes matter for some of Spark's most biting comedy; but no less than in its predecessor are we in a fallen, and not merely an imperfect, world. In both novels the situation would be unbearable if the deliberate avoidance of emotion did not transform each book into a puppet-show, a morality piece. What saves them both from morbidity is Spark's agile humour, gift for dialogue, and grasp on the material world. We are not simply in the realm of private fantasy or obsession.

In *The Driver's Seat* (1970) that realm of private obsession does take over, though as itself the subject of a story in which the element of play becomes the vehicle for an examination of the nature of predestination. With its unnerving portrait of the convergence of murderer and victim (which is which?) this novella has a Calvinstic sense of inevitability. The handling of the material is anything but laboured. The narrative is couched in the historic present: this is a *fiction*, a portrayal of the way things happen. Ostensibly there is nothing supernatural inherent in the theme; but the implications of the selection by an unbalanced woman of the man she knows can kill her take on a parabolic character. It is not wrongdoing, it is evil, which is in evidence. And Spark's balancing of present and future tenses, the former depicting Lise's choices, and the latter employed to pronounce their inevitable fulfilment, enacts in syntactical terms the operation of human volition in the context of God's foreknowledge. Lise's own state of mind, her boredom and materialistic self-concern, is itself a portrait of the Godless despair endemic in that commercialization of human life to which the Western world submits itself and which it spreads. Spark's refusal to heighten the narrative voice by any eerie or dramatic effects only serves to make the tale the more convincing through its banal details.

She reverses the process in *The Hothouse by the East River* (1973), which like so many of her novels begins with a conceit, in this case the notion of a woman whose shadow falls without reference to the sun. A writer such as Blackwood might have developed the image into a story about someone living in a different dimension to other people; the emphasis would have been on the paranormal or the parabolic. But Spark plays around with the idea, fantasizes without logic, and ends up by revealing that the woman and her husband are both dead. On a first reading the effect is capricious and nonsensical. This is not a matter of the concept as such: Charles Williams, for instance, more than once portrays the interaction of the dead with the living within a providential network of redemption. By combining the lives of a dead man and woman with those of the children they could never have had, Spark juggles past and future to produce a multidimensional universe that is itself a fantasy. Even more than in Williams's novels, we are immersed in an eternal present. Spark simply abolishes any serious feelings of chronology by placing her characters in relationships to which such concepts as past and future are of limited relevance. The excessive heat of the apartment overlooking New York's East River, where Elsa lives, suggests a purgatorial context, as does her obsessive turning towards the east—in the hope of the coming of her Saviour? But all interpretations of a fixed kind are doomed to be unsettled: this novel rejects all systematic readings. Spark's use of the historic present has the effect of securing a timeless abstraction from chronology; and in *The Hothouse* this has the effect in turn of reinforcing the relativity of fictive truth.

She casts a shadow behind her as she moves her chair to make room for him. Today she began a new course of analysis, or perhaps she began last week.
  She is saying, 'I bought a pair of shoes.'
  Or, 'Pierre doesn't know what to do.' (ch. 1)

Elsa and the other characters are in a fictional limbo: none of their actions or words are necessary, any more than the author's decisions as to their movements are unchangeable. As these characters mix with others, who either exist independently as do (possibly) their two psychoanalysts or (even) their two children, the absolute relativity of the material world itself seems called in question. But the effect is as much of something surreal as of something supernatural.

This weird perspective reinforces the undercurrent of terror that runs through all Spark's finest work. It is most eruptive in *The Driver's Seat*. At the moment of her self-sought death Lise has to confront her own experience of the ultimate. 'As the knife descends to her throat she screams, evidently perceiving how final is reality.' The ambiguity of 'evidently', the contextual transference of meaning from 'presumably' to 'certainly' only adds to the moment's force. And the concluding words of the book embody the finality of horror as the hapless killer 'sees already the gleaming buttons of the policemen's uniforms, hears the cold and the confiding, the hot and the barking voices, sees already the holsters and epaulettes and all those trappings devised to protect them from the indecent exposure of fear and pity, pity and fear'.

Muriel Spark is the late twentieth-century English supernatur-alist *par excellence*, her work free from any confusion between the supernatural and the paranormal: the former is invariably tran-scendent and unknowable, and not to be contacted or deployed. The mysterium is for her the great unsayable and thus the great unsaid; and this saves her from assaulting it with adjectives, from trying to take it by the storm of magic. Indeed, magic is as suspect for her as it is for Murdoch, and she portrays a fine range of seedy, bogus supernaturalists, all of whose methods and pre-occupations are irrelevant to the life of fact and the fact of life. And just as in *Memento Mori* death is presented as a fact of life, so in *The Hothouse by the East River* life is shown to be a fact of death. Nor are moral categories predictable. In *The Girls of Slender Means* (1963) a vision of pure evil occasions a religious conversion —everything that exists is a manifestation of the mysterium. If God be God then he is indeed eternal, omnipresent, and ines-capable. Everything is relative to his absolute being. At root the vision is as direct and simple as that of Anne Brontë.

Spark's novels are as obviously works of artifice as are those of William Golding; she herself describes them as 'a pack of lies'.[14] Avoiding the pretensions of ultra-naturalism, she insists on the creative force of imaginative actions, on the self-generating powers of language. Her work proffers only implicit evidence of things not seen, for

it is better to know what is doubtful than to place faith in uncertainties. Doubt is the prerogative of the believer; the unbeliever cannot know doubt. And what is doubtful we should doubt well. But in whatever touches the human spirit, it is better to believe everything than nothing. Have faith.[15]

Muriel Spark's first novel coincided with the penultimate one by Evelyn Waugh (1903–66). Writing to the novelist Gabriel Field-ing, who had sent him a copy of *The Comforters*, Waugh com-ments, 'The mechanics of the hallucinations are well managed. These particularly interested me as I am myself engaged on a similar subject.'[16] He then adds with obliging acerbity, 'Mrs Spark no doubt wants a phrase to quote on the wrapper and in advertisments. She can report me as saying: "brilliantly original and fascinating".' The burlesque of such promotional tributes goes sounding off as autonomously as do the disembodied voices that incubate in both *The Comforters* and *The Ordeal of Gilbert Pinfold* (1957).

The fact that the fictitious Pinfold is an obvious self-portrait and that his delusions happened to Waugh himself imbues the book with a biographical interest that distracts from its real achievement as a piece of fiction about the self-delusive power of fiction. The voices Pinfold hears resemble a novel that has become a text without authorial control. At the end he sits down to write the novel we have just been reading, imposing order through the deliberate exercise of solipsism. Fictional 'imaginary' experience thus becomes fictive imagined experience; material delusion becomes imaginative truth.

Initially Pinfold's experience is puzzling (where do the voices come from?) and the effect is preternatural, invasive, both in nature and in context an external threat. He confronts the variously abusive, conspiratorial, ribald, and insinuating voices by rationalizing them; and in the process Waugh neatly satirizes a materialistic mind's attempt to cope with the inexplicable.

He supposed that somewhere over his head, in the ventilation shaft probably, there were a number of frayed and partly disconnected wires which every now and then with the movement of the ship came into contact and so established communication now with one, now with another part of the ship. (ch. 3)

The hope behind the lazy vagueness is that the preternatural, once it has become paranormal, can be mastered.

The voices are only finally expelled by Pinfold's own resolution: he accepts responsibility and performs an act of exorcism through what amounts to an assertion of objective fact, the fact that they bore him and make him angry. Nevertheless, the book ends on a note of mystery. While what the voices say can be attributed to Pinfold's own recent past and to his character, temperament, and occupation, the reason for their eruption is not entirely covered by the fact that he has been accidentally poisoned. They are to be accounted for neither medically nor psychologically. Pinfold complains, 'If I wanted to draw up an indictment of myself, I could make a far blacker and more plausible case than they did,

I can't understand.' The author then comments that 'Mr Pinfold never has understood this; nor has anyone been able to suggest a satisfactory explanation' (ch. 8). The remark recalls the Commandant's observation at the end of Golding's *Free Fall*, 'The Herr Doktor does not know about peoples.' Human nature necessarily involves encounter with imponderables. As Pinfold settles down to write his book he is on the threshold of the mysterium.

All the novelists discussed in this chapter were writing in an age to which the concept of the supernatural had become alien or at least eccentric. Moreover, the ever-increasing comprehensiveness of literary naturalism had as its corollary an extension of materialistic assumptions as to the sources and experience of human consciousness: it is the paranormal rather than the supernatural which provides the expository vocabulary for any manifestation of the mysterium. However, as a further corollary, the extended nature of the subjects open to naturalistic treatment allows for the inclusion of abnormal, inexplicable, and purely psychic happenings which arise in the field of the emotions and above all in that of sex. Accordingly, subjects once taboo and written off as 'supernatural'—that is to say, 'irrelevant'—return through the back door of a more searching realism.

In the writings of Fowles, Golding, and Murdoch the irrational is not rationalized but accepted and explored as a means of extending the boundaries of human awareness and moral choice. In the case of the last two, choice very evidently involves responsibility, and the deliberate artifice with which supernaturalist motifs are deployed by them itself furthers one's sense that for these writers—for Murdoch especially—human beings exist alone before a mysterium which both governs and judges them. For her, the supernatural, allowing for the eccentric phenomenology in her fictions, is located in morality, the mysterious ubiquity and imperative of good.

While the same thing could be said of the writings of Spark and Golding, the deliberate use of fictive play in their novels is far more prominent. In Golding's case the result is that many of his novels take on the nature of fables, of the specifically didactic and expository. In Spark's case the ludic elements are quite detached from any ostensible moral comment. Her novels are the very apotheosis of literary irony: their readers are challenged to look for an elusive meta-text whose presence is obliquely intimated by the outlandish (not to say preternatural) nature of the fictional material. By employing supernaturalist motifs in the teeth of, yet in terms of, contemporary materialism, this brilliant player of literary god-games leaves the ontological exploration of the supernatural as open as before.

9
REVERSIONS TO TYPE

Imagination has nothing to do with Memory.

(William Blake)

'Last night I dreamt I went to Manderley again.' There could be no greater contrast with the opening of *Agnes Grey* than the insidiously melodious sentence that conducts the reader into Daphne du Maurier's *Rebecca* (1938). All the same, with its emphasis on dreaming rather than on instruction, that opening reveals its author's affinities with those earlier romance writers who had inspired the two elder Brontë sisters to compose their own more serious and intellectually demanding fictions. As it happens, however, the opening of *Rebecca* springs a trap: the lilting cadence seems to promise something nostalgic and consolatory,[1] whereas what Manderley turns out to represent is terror, destruction, and hostility. Romantic though it is, the book is not a supernaturalist novel, though the author understood the dual appeal of the supernatural to post-Victorian readers—that it should either disturb and frighten, or provide comforting escape from more pressing everyday concerns. The two tendencies were to continue past du Maurier's day up to the present time, though the dichotomy between those contrasting requirements has not been productive—and so long as the dichotomy is insisted upon, could not be productive—of major art from writers working at either extreme of the divide.

For despite such pleasurable imaginative alternatives, traditional embodiments of the unseen have become sources of embarrassment; they need more than nostalgia to keep them alive, for nostalgia is the preserve of secondary fictions. A liberating disillusionment leads to the destruction of false gods; and if late twentieth-century novelists opted out of simple naturalism into a more deliberate fictiveness, it was presumably because such arbitrary uses of the imagination enhanced authorial responsibility. Although the concept of the supernatural as divine order was no longer a matter either of proof or refutation, for imaginative writers it continued to be one of choice.

The Loss of Paradise

The second half of the twentieth century in England has therefore been both the best of times and the worst of times for authors wishing to write about the supernatural. Their advantage has been that the break-up of naturalistic realism into a variety of literary methodologies has enabled them to introduce such eccentric themes without danger of being labelled specialist, arcane, or quirky; on the other hand, the defensive scepticism which attends most sophisticated confrontations with the notion of the unseen has proved to be emotionally crippling. Positive supernaturalism, the intellectual commitment to a metaphysical world-view, has seldom been found among the major writers of fiction; but at the same time the negative supernaturalism of threat and terror has been superseded both by the dreads engendered in the individual subconscious and by the random chaotic violence let loose with the collapse of the traditional social and moral structures of industrial civilization. The title story of Elizabeth Jane Howard's *Mr Wrong* (1975) therefore presents an authentic contemporary image of the individual's fear of exposure and incapacity, in the shape of a haunted motor car. Its account of night driving, of the world of service stations and casual pick-ups, carries a suggestive horror that reflects the peculiar lonely ghastliness of the latter-day moral climate of the roads: in this instance the preternatural is well-nigh squeezed out by material terrors. Indeed, the problem nowadays for the would-be purveyor of supernatural dread is that the modern fragmented, mechanized world can itself be so frightening to solitary vulnerability as to preclude any genuine dread of the supernatural. It is humankind, not God, which raises devils.

It is therefore no wonder that writers continue to dream of a lost paradise, of an Eden still recoverable, could one but find the key to it. Supernaturalist fiction has its reassuring, beneficent aspects as well, maintaining a vision of harmony between humanity and the world of nature, between the invisible and what is seen. It is a comment on the general fading of such conviction that it should be voiced most often in writing designed for children and in the gentler kind of fantasy, a genre that encourages a domesticated supernaturalism. Thus Kenneth Grahame's animal fable *The Wind in the Willows* (1908) presents the figure of Pan as 'The Piper at the Gates of Dawn', a gracious figure in an idyllic riverside setting. Very different from the phallic demons of Saki or Arthur Machen, he is the object not of terror but of romantic reverence in a world that is fundamentally safe and sound. But it is to animals not to human beings that the epiphany occurs.

As a purveyor of a gentle paganism Grahame's most significant successor is the Ulster writer, Forrest Reid (1875–1947). Reid's imaginative preoccupations were with the world of boyhood—and with adolescence: in the words of his friend and admirer E. M. Forster, 'he was not interested in maturity'.[2] It sounds a damning indictment, but Reid himself would not have been dismayed by it, as his account of his own boyhood, *Apostate* (1926), makes quite clear. He describes himself as an orientalized pagan, and voices the early twentieth-century ideal of a quasi-mystical harmony with nature.

There were hours . . . when I could hear the low breathing of the earth, when the colour and smell of it were so close to me that I seemed to lose consciousness of any separate existence. Then, one single emotion animated all things, one heart beat throughout the universe, and the mother and all her children were united.[3]

This is the same sense of unity that one finds expressed with similar fervour in Jefferies's *The Story of My Heart*; but the mother-child image personalizes and softens the conception in a way peculiar to Reid.

His novels are filled with a sense of the impending presence of 'a spiritual world much closer to this world than the remote heaven of Christianity',[4] one mediated both through nostalgia and a sense of danger:

What was this sudden fever of restlessness crying like an ancient cry within him? A melancholy crept over him, and the strange unearthly beauty of his surroundings grew sad too. It was as if the whole scene had retreated from him, as if he were no longer in it as part of it, but only as a stranger from another world. So a ghost might feel whose hauntings were unperceived and unsuspected.[5]

This sense of sadness is aroused by an overgrown garden, a pool, and the statue of a naked boy, such a scene as Walter de la Mare also frequently portrays. But there is an eroticism in Reid's work which is entirely lacking in de la Mare's.

Reid's most ambitious supernaturalist novel is *Uncle Stephen* (1931). An imaginative 15-year-old, Tom Barber (whose earlier boyhood Reid was to describe in *The Retreat* (1936) and *Young Tom* (1944)) runs away from his unloving step-family to a hitherto unknown great-uncle, a former wanderer now turned recluse. The pair establish an immediate rapport—one which the two later books present as existing in a timeless spiritual dimension. In these we learn of Tom's psychic gifts (he sees a ghost and sparks off a poltergeist disturbance; he talks to animals and they reply to him; he is haunted by dreams of a malevolent old magician who is a presage of what Uncle Stephen could, but never does, become). The uncle is a student of the occult, and allows himself to revert to boyhood in order to be Tom's companion, an adventure which threatens to entrap him in his past (a possible warning by the author to himself, concerning the dangers of his own preoccupations). The boy Stephen, alien yet defiant in what to him is a future world, is a convincing image of the child within the man who continues to exist into adult life, who psychically is never lost. However, the novel is clear as to the dangers, as well as the attractions, of clinging to the past and to lost innocence. Reid handles the metaphysical implications with a deft touch. Far from being a whimsical frolic in the space-time mode, *Uncle Stephen* ponders the nature of timelessness and endorses the saving grace of the present moment and the experience of the physical life of natural objects and real people.

Reid's novels offer their own challenges to literal-minded readers. Their very guilelessness in recounting the affectionate comradeship of boys is as much of its time as are the endearments in Forster's *Maurice*, Henry Williamson's *The Beautiful Years* (1921), and the tales and fantasies of Richard Middleton. Were it not for the supernaturalist element, which the author projects through Tom's youthful consciousness, the books might be dismissed as mere pederastic reveries; and late twentieth-century readers are as likely to be as flustered by these accounts of physical tenderness as was Henry James by Reid's unannounced dedication to him of an idyll of schoolboy love, *The Garden God* (1905). In the Tom Barber novels, at any rate, this particular sensibility is offset by an occasional level-headed tartness that modulates the more plangent, erotically charged exchanges. It is the beauty of supernaturalism, and the human responsibility for interpreting its presence, which determine the course of the trilogy. Reid is less interested in the nature of time than in the quality of eternity, not so much concerned to describe emergent sexuality as to capture the lingering emanations of a spiritual consciousness dormant in the life of reverie and dream.

The Ageing of Innocence

It could be argued that Reid's vision is too circumscribed by his idolization of youthful loves and dreams for it to have more than a personal, subjective relevance; and following the Second World War nostalgia for a lost world order began to be replaced by hopes that it was within human capacities to build a new one. Paradise could be in the making. Children's fiction became more consciously related to social issues; it began to take account of the changes in popular consciousness. It also grew more self-conscious of itself as a genre, providing its authors with a literary challenge, especially if they had a cause to espouse. One obvious example of this use of children's literature for didactic ends is in the widely read Narnia series of C. S. Lewis; and a still more methodical and subtly structured example of the kind is 'The Dark is Rising' sequence of Susan Cooper, which was published between 1965 and 1977 in the heyday of the genre. These five books appeared in the post-Tolkien era, one in which there was much popular interest in archaeology, ley-lines, and the Matter of Britain. Cooper's sequence capitalizes on all these enthusiasms and adapts them skilfully as reading material for young adolescents; but it also grows in complexity and richness, and like T. H. White's *The Once and Future King* (1958), at a literary level it matures with its protagonists.

The books are unusual in providing their own metaphysic. 'This where we live is a world of men, ordinary men, and although in it there is the Old Magic of the earth, and the Wild Magic of living things, it is men who control what the world shall be like.' The speaker, an 11-year-old boy who is one of the Old Ones, guardians of the supernatural mysteries, is talking to his grown-up brother. The stilted and formal nature of the language may arise from its need to be plain enough for a child to use and understand, yet not so childish as to be reductive, or alien from an adult reader.

> But beyond the world is the universe, bound by the law of the High Magic, as every universe must be. And beneath the High Magic are two . . . poles . . . that we call the Dark and the Light. No other power orders them. They merely exist. (*Silver on the Tree*, ch. 1, author's ellipses)

The sequence is interesting for its eschewal of any avowedly Christian content; yet the dualism of light and dark is itself subject both to the High Magic and to a more familiar humanism. The battle between dark and light leaves humans responsible for their own fate: Cooper depicts it as operating outside of time, so as to provide the background for human choice and conflict. But it in the light which is in final control. The High Magic may or may not correspond to what traditional piety knows as God, but the division between light and dark is absolute: it is no question of the Yin and the Yang in which two categories each contain an element of the other's being, so that in due course they exchange roles and natures. As the Welsh farm worker John Rowlands puts it in *The Grey King* (1975), 'At the centre of the Light there is a cold white flame, just as at the centre of the Dark there is a great black pit bottomless as the Universe.'

This metaphysical structure gives the five books their particular quality; and it also keeps them from becoming sentimental. The Light is no familiar, friendly category, and is not assumed into the author's private imaginative world as it is, for instance, in Lewis's *That Hideous Strength*:

> those men who know anything at all about the Light also know that there is a fierceness to its power, like the bare sword of the law, or the white burning of the sun . . . At the very heart, that is. Other things, like humanity, and mercy, and charity, that most good men hold more precious than all else, they do not come first for the Light. (*The Grey King*, 'The Pleasant Lake')

It is an austere vision to be presented in a book for children, and faces up squarely to the paradox endemic in humanitarian liberalism, the paradox that compassion appears to involve a tolerance of evil. The definition comes near to an understanding of the supernatural nature of the good in absolute terms; or rather, it displays what the claims for such an absolutism can imply.

The conflict between Light and Dark is fought out through a wide variety of motifs and through many familiar myths and symbols. The Arthuriad, with Merlin embodied in the figure of Merriman, an 'uncle' to three of the child protagonists; the Wild Hunt; Bran the Blessed and the land lost beneath the sea; rural demonology; time-travel; Owen Glendower and the Welsh historical mythos; ritual magic, dreams, and second sight—it is a heady mixture. Cooper also makes use of familiar landscapes out of literature and legend—the peaceful Thames Valley of *The Wind in the Willows*, a Cornish fishing village from the worlds of Arthur Quiller-Couch or Daphne du Maurier, and the *Mabinogion* country of Mid-Wales. The story itself is multi-dimensional, operating in several times at once; and in *Silver on the Tree* (1977), the final volume, events proliferate with a speed that suggests the helter-skelter quality of John Masefield's books for children, *The Midnight Folk* (1927) and *The Box of Delights* (1935). But Cooper employs a sterner intellectual control than does the perennially boyish Masefield: she is never merely playful and is indeed rather humourless. *The Dark is Rising* (1966), with its description of a family Christmas, comes the closest to conveying a natural warmth; even so, that Christmas is infiltrated by terrors that make this the most intense and original of the five novels. It is appropriate that it should provide its title for the sequence.

Susan Cooper's is a thoroughly worked-out achievement, and crafted with an almost technological efficiency. There is great deliberation behind the deployment of the multifarious happenings; and it is this, perhaps, which accounts for the pervasive hardness of tone, a clear-cut knowledge of what the story is about, which keeps it from attaining the depths of genuine mystery; skilful and intellectually stimulating though much of the sequence is, the authorial control prevents any sense that truth is being discovered in the act of writing. There is an impression of computer printout about the inclusion of so many proven magical ingredients, just as the quest for the various sacred objects—the Grail, the harp of gold, the various rings of power—feels rather like a treasure hunt, with the various emblems swiftly discovered after the appearance of each enigmatic clue. The whole sequence, vigorous and exciting though it is, seems to mark the end of a tradition. The metaphysical structure guarantees the naturalism, not the other way around. Innocence is lost.

Loss of innocence is a factor that faces any contemporary writer who deals with the supernatural as it impinges upon young people: the kind of children who figure in the pages of Grahame,

Reid, and Blackwood are, it would seem, no more. Rather it is
the vulnerability of adolescents exposed to the physical and psychic strains of their emergence into adult life which provides the
most natural subject on which the supernatural can be seen to
impinge, to either creative or destructive effect: personality and
passions are here at a point of change. It is for this reason that
a novel like John Gordon's *The House on the Brink* (1970) lends
itself with such effectiveness to a supernaturalist theme. Set in
and around a Fenland town, it makes good use of the peculiar-
ities of local atmosphere and landscape such as its young protagonists would know intimately and by which they have been
formed. The portrayal of the malevolent energy of fear and superstition, and their power to dredge up destructive elemental
forces, combines a sense of preternatural menace with a robust
and companionable understanding of the adolescent lovers, who
become conductors for emotions and events that open upon
dimensions which fascinate and yet repel them. The story hovers
between two worlds of experience, that of vulnerable and shackled
age, and that of a youthfulness insecure in its very possession of
hope and strength. It can speak not only to younger readers but
to the elderly, who are in a position to appreciate the ironies and
terrors it explores. The appeal of the supernatural has always
crossed the barrier of age.

Gordon's novel has, in common with much of the best late
twentieth-century supernaturalist writing, a tendency to earth its
spiritual happenings in the world of everyday, a tendency to move
from the magical to the sacramental. One can see this process
at work in the very development of the fiction of Alan Garner (b.
1934). An inventive writer for children, he is at the same time a
regional novelist who writes for adults. His work reveals a growth
from an uneasy use of preternatural incidents, through a concern
with the paranormal, into a supernaturalist perspective on the
natural. In *The Weirdstone of Brisingamen* (1960) and its sequel
two children become involved with such mythical beings as the
Wild Huntsmen and a shape-shifting darkness, the Brollachan,
an elemental out of Highland folklore. The young protagonists
being mainly eyes and ears, one is compelled to join in the
adventures, rather than to have one's attention focused on the
characters experiencing them. The four chapters in *The Weirdstone*
which describe a journey through underground tunnels have a
physical immediacy that seems to cry out for inclusion in a different kind of book.

Elidor (1965) begins to shift the perspective from the mythological to the everyday: the novel describes with ingenious particularity the interaction of two worlds. The stricken wasteland of Elidor is only once, and that briefly, visited by the four children, whose task it is to safeguard in their own world (itself an urban wasteland) the four sacred emblems of Celtic mythology from attack by the dark powers. In Elidor these emblems are a spear, a sword, a stone, and a bowl, which in the children's world of suburban Manchester are converted into 'a length of iron railing . . . a keystone from the church . . . two splintered laths nailed together . . . an old cracked cup, with a beaded pattern moulded on the rim'. The attempts of the enemies of Elidor to break through into the children's world are described in paranormal terms, for the sacred emblems generate a static electricity which interferes with television and starts up a car battery: Garner's collocation of preternatural modes is more logical than is Constance Holme's in *The Old Road from Spain*, the power of Elidor being expressed 'not in magic but in physics'.[6] The final breakthrough is described in language that embodies to the full Garner's ability to describe a paranormal happening.

A white spot had appeared in the middle of each shadow, quivering like a focused beam of light. The spots grew, lessened their intensity, changed, congealed, and became the expanding forms of two men, rigid as dolls, hurtling towards the children. They matched the outlines of the shadows, and were rising like bubbles to the surface. As they came nearer their speed increased: they rushed upon the children, and filled the shadows, and eclipsed them—and at that instant they lost their woodenness and stepped, two men of Elidor, into the garden. (ch. 16)

The precision of this makes it read like something actually observed; yet because of the one-sidedness of the presentation the two worlds coalesce uneasily. Garner deploys one kind of language for Elidor, another for the world of Manchester, without any sense of overlapping realities. No more than the preternatural mythological mode is the paranormal one adequate to his particular vision.

In *The Owl Service* (1967) he moves into the stressful world of adolescence, and tackles a mythological theme in a more systematic and convincing manner. This story of the re-enactment in a remote Welsh valley of a legend from the *Mabinogion* is told in a terse, elliptical style entirely appropriate to the emotions of the three adolescents who find themselves caught up in the power of the ancient drama which those very emotions have themselves served to unleash: the girl Alison becomes a natural conductor, and the rivalry for her sympathy between her stepbrother Roger and the working-class Welsh boy Gwyn re-enacts that of Llew Llaw Gyffes and Gron Pebyr in the legend.[7] Garner's interpretation of the myth elaborates upon its implications: Blodeuwedd, created out of flowers, is changed by her vengeful creator Gwydion into an owl, so that when her pent-up energies are released she inevitably goes hunting. In the words of the gardener Huw, the mouthpiece of Gwydion, 'When I took the powers of the oak and the broom and the meadowsweet, and made them woman, that was a great wrong—to give those powers a thinking mind!' And it is Huw who voices the tragedy. 'She wants to be flowers . . . but you make her owls. Why do we destroy ourselves?' In *The Owl Service*, as in *The Old Road from Spain*, the eternal recurrence is broken, but not in this case by death: it comes through the realization that Blodeuwedd's true nature is not birds but flowers. The novel closes with a recovery of original innocence and joy.

The strong point of *The Owl Service* as a tale of the supernatural is that the preternatural happenings are perfectly integrated into the daily life of the valley, as when the sheepdogs chase Gwyn down from the mountain and so prevent his running away. The human dimension likewise accords with the preternatural events: as Gwyn says, what fills the valley is not a haunting from the past but something that is still happening. The same comment applies to *Red Shift* (1975), in which Garner presents three stories set in different epochs which take place in the same landscape. The tales are told concurrently, and concern a pair of contemporary teenage lovers, a group of second-century tribesmen, and a number of village people at the time of the Civil War. Linking them is a stone-age axe-head; and occasionally the stories seem to overlap. Told in a bare, allusive language akin to that of William Golding, all three episodes are raw with a sense of bitterness and pain. This simultaneity gives little or no sense of providential care. Timelessness is seen as a scientific rather than as a religious fact.

This point of view reaches its logical conclusion in the four short tales that have come to be known as *The Stone Book Quartet* (1976–8). Nothing preternatural occurs in them, yet they provide a sense of integrated mystery, expressed through the portrayal of traditional crafts as being themselves a mystery. For these tales amount to a hymn to good craftsmanship and to the wholeness of the experience of those who work with their hands and form part of traditional customs and live in long-settled communities. 'The single quality which most distinguishes the books is the specificity of their imagery, their rooting in a physical world of ploughs, looms, hammers and forges; even the most metaphysical passages are expressed in terms of clock or weather-vane.'[8] The writing is as terse and allusive as in *Red Shift*, but often with the dramatic immediacy of the early stories: few people who have read the account in *The Stone Book* (1976) of Mary's ascent of the ladder up the church spire are likely to forget it. Still more memorable is Garner's presentation of interlocking times, and the mediating power between the generations of cherished objects such as a clay pipe or the two horseshoes which the boy William appropriates at the end of *Tom Fobble's Day* (1977). In this story, and in connection with the horseshoes, we find the quintessence of Garner's vision. Williams's grandfather Joseph declares that the shoes hidden up the chimney are his greatest treasure. 'They're our wedding . . . They're us!'

Your friends and your neighbours give them to the wedding. No one says. It happens. And it happens as the smith's at his forge one night, and happens to find the money by the door. And he makes the shoes alone, swage block and anvil: and we put them in the chimney piece. Mind you, I'd know Tommy Latham's work anywhere. But we don't let on. It's all a mystery.

The puzzle element (the logistics of the horseshoes' provenance) is fused with the element of mystery, the word being used here in its fullest range of meanings. This is the world Kipling describes in the Puck stories: the magic is a natural magic, born of *pietas*, goodwill, and reverence for life.

Below the Surfaces

Garner's achievement is hardly new: it is its use of a sophisticated literary technique to attain its end which makes it different. The end-vision recorded in *The Stone Book Quartet* recalls that of the Scottish writer Neil M. Gunn, whose early novels, such as *Morning Tide* (1931) and *Highland River* (1934), devotedly record the lives of fishermen and crofters and the lingering traditions of the glens.[9] The later novels develop a preoccupation with supernaturalist themes, *The Silver Bough* (1948) being representative in incorporating them into a naturalistic story. An archaeologist unearths two kinds of mystery. One is a crock of gold, the stuff of legend and romantic mystification—ultimately a puzzle: as such it is appropriately purloined and hidden by a simple-minded peasant, a throw-back to the aboriginals who placed it in the tomb. But with it are disinterred the skeleton of a mother with her child, the ultimate human mystery, life itself, inexplicable but reincarnate in the persons of a local girl and her bastard daughter. This novel constitutes a vision of the coinherence of past and present through an intricate network of references and re-enactments: it celebrates a richness and density in life which the author sees as being completely overlooked by the irrelevant simplifications of the scientific and materialistic points of view.

The ability of writers like Neil Gunn and Alan Garner to reanimate the contemporary sense of mystery, by exploring the continuous indwelling of past cultures and beliefs, springs from an authentic response to the immanence of the supernatural in the material world, in Garner's case mediated through an apparent dissatisfaction with conventional iconography and literary traditions. That that dissatisfaction is justified can be demonstrated by their use in what have been two of the more esteemed supernaturalist novels of their time, both of them by authors who had made their reputations in other fields. Kingsley Amis's *The Green Man* (1969) is particularly interesting as an example of a supernaturalist novel written within a realist, satirical tradition. The narrator, Maurice Allington, the kind of bibulous womanizer who is a staple of the author's fictional world, runs The Green Man, a pub named after a wood-demon once raised by the seventeenth-century sorcerer who still infests the place. The latter fastens on the vulnerable Allington and lures him into reawakening the green man. So far the book might be interpreted parabolically; but Amis seems to grow impatient with his subject-matter. Having provided a theophany whereby God appears to Allington in the shape of a dandified young man who admits that the creation of the human race is a game He has thought up for His own purposes (an orthodox idea in an unorthodox guise), he then has the sorcerer's spirit reluctantly exorcized by a disbelieving liberal clergyman, whose portrayal, forged in the smithy of a rancorous prejudice, is so distorted as to subvert any serious intent the book may have.

At one level *The Green Man* addresses the nature of death and

the possibility of immortality; but Amis does not work happily in
the field of the supernatural. The preternaturalist effects come
straight out of the pages of M. R. James, and so unemphatically
as to appear offhand; while the theophany, for all its intelligent
and even witty dialogue, is conducted as though it too were merely
preternatural. It constitutes a literary parlour trick, couched in
the same sub-acid yet jaunty tone as the rest of the novel. *The
Green Man* is less interested in metaphysics than in morality. Its
use of preternaturalist machinery is directed towards Allington's
own state of mind and to his effect upon his family rather than
towards the nature of reality in any wider sense. The ghostly
manifestations are put to use for domestic ends, being subject to
the narrator's own conviction that a phantom can harm no one,
and is as nothing to the devilish powers with which the mind can
torment itself.

Such a conclusion is scarcely in evidence in Susan Hill's *The
Woman in Black* (1983). Like Amis, Hill is a writer who should
be of interest to future historians of the novel, having produced
examples of several familiar and popular types of fiction—the
domestic novel, the rural novel, the post-Freudian novel of childhood, the war novel, and so on, each of them written with the
assured and economic ease of one who has read widely and
learnt her literary skills in the process. Her range is as broad as
that of, in his day, Edward Bulwer-Lytton, and like him she was
to produce as one of her most effective narratives a novel concerning the supernatural. A self-styled 'ghost story', *The Woman
in Black* (the reference to Wilkie Collins's novel is surely deliberate) is set in the heyday of the genre, the early twentieth century,
and contains a number of familiar and well-seasoned ingredients—the solitary haunted house; the reserved local people who
know more than they will tell; the dog with psychic responses;
the silent figure of the woman in black, baleful as Miss Jessel as
she confronts the narrator in an abandoned graveyard; the
narrator himself, the young London solicitor stumbling to his
undoing among family secrets still manifest in a child's screams
at night, a locked room, and a phantom horse and cart upon the
marshes. An entire literary tradition is distilled with a sure sense
of its imaginative potential. But the novel is rather more than
pastiche. What gives it its distinctively late twentieth-century
quality is its refusal of any sense of providential oversight. The
evil it portrays is both visible and potent: the mystery of pain and

injustice goes deep, and human resentment of the mystery makes it go deeper still. *The Woman in Black* embodies a vision of the tragic working of evil, operative within the eternal present. It lacks the remedial optimism of *The Green Man* and demands a more comprehensive theophany than that book provides. Powerful though it is, however, it does stand at the end of a tradition. Susan Hill's very success in distilling the essence of this particular literary medium may well have exhausted its future possibilities.

Yet supernaturalism continues to be a force among contemporary writers, most notably perhaps in the work of Peter Ackroyd, in several of whose fictions the interleaving of past and present is carried out with striking virtuosity. *Hawksmoor* (1985) could well come to be regarded as the quintessential supernaturalist novel of its time, one which sums up many of the themes discussed in the foregoing pages. In the first place, it is a powerful tale of terror that also makes use of hermetic material; and it both internalizes the element of preternatural horror and conveys a numinous sense of place. That place is the East End of London, and Ackroyd's evocations recall the best work of Dickens, Stevenson, and Arthur Machen. For *Hawksmoor* is at one level a murder mystery and at another a story which presents a spatial view of time.

Its central character exists simultaneously in the early eighteenth century and in the later twentieth. As Nicholas Dyer he is supposed to be the pupil of Christopher Wren, the colleague of Vanbrugh, and the designer of the six great London churches really built by Wren's actual pupil, Nicholas Hawksmoor. In the novel, however, the latter name is borne by a twentieth-century detective who is trying to solve a series of murders in the grounds or crypts of the various churches, murders which are manifestations or re-enactments of ritual slayings carried out by Dyer in his own century. To him the world is in the hands of the Lord of Darkness. He scorns the scientific enlightenment of Wren and the Fellows of the Royal Society, and some of the most stimulating passages in the book are formed by his disputes with Wren, in whose company he visits Stonehenge: *Hawksmoor* echoes *Tess of the D'Urbervilles* and *A Glastonbury Romance* in making use of the great stone circle as an emblem of timeless mystery. Wren is interested chiefly in its scientific measurements and purpose; yet while there he receives an intimation of the death of his son thousands of miles away.

This contrast of two world-views is at the heart of the book, which appears to endorse Dyer's reading of experience at the expense of Wren's. Orphaned in the Great Plague and befriended by a diabolist, the former is well aware of the pagan roots even of Christian worship. 'I address myself to Mysteries infinitely more Sacred and, in Confederacy with the Guardian Spirits of the Earth, I place Stone upon Stone in Spittle-Fields, in Limehouse and in Wapping' (ch. 1). Each church is to have a ritual victim buried beneath it. It is in the religious sense of the term that one of the boy victims is to recall that at school his fellows would refer to history as 'mystery'.

Ackroyd is a master of pastiche, and his evocation of the London of Queen Anne is painfully convincing. The physical squalors of the life of the very poor suffuse the book; so too does its savagery. Dyer reflects that

Destruction is like a snow-ball rolled down a Hill, for its Bulk encreases by its own swiftness and thus Disorder spreads; when the woman nam'd Maggot was hanged in chains by here, one hundred were crushed to Death in the Tumult that came to stare upon her. And so when the Cartesians and the New Philosophers speak of their Experiments, saying that they are serviceable to the Quiet and Peace of Man's life, it is a great Lie: there has been no Quiet and there will be No peace. (ch. 5)

The two centuries are linked together by recurring imagery, so that when Hawksmoor visits his father in an old people's home the scene virtually reproduces that in which Dyer and Wren visit the lunatics in Bedlam. Further connections are made through the constant appearance of small children singing nursery rhymes and street songs, often with gruesome implication; of copulating couples caught in the act; of vagrants and of frightened cats: characters reproduce themselves, time is engulfed in space. Above all, the churches become powerful images of mystery. Ackroyd conjures up the peculiarly startling effect these eccentric buildings can have on a spectator. One can readily accept Dyer's contention that 'I build my churches firmly on this Dunghill Earth and with a full conception of Degenerated Nature' (ch. 1).

Dyer himself recalls Melmoth the Wanderer: his bitter misanthropy is itself a necessary challenge to an otherwise superficial rationality. The book is full of an angry pity for the fate and condition of men and women in the world—pity where the boy victims are concerned, anger over the follies and self-indulgence of adult humankind. The sense of evil is in places overpowering: Ackroyd can hold his own with Le Fanu as a master of the threatening and macabre; and his novel has a detailed naturalism forbidden by literary convention to his predecessors. The spiritual terrors are reinforced by a physical immediacy: the reading process here becomes as much a determining factor in the experience of the novel as it does in the more cerebral work of Spark or Golding.

Not least among the interesting features of *Hawksmoor* is that it provides a virtual apologia for the supernaturalist novel as such.

Men that are fixed upon *matter, experiment, secondary causes* and the like have forgot there is such a thing in the world which they cannot see nor touch nor measure: it is the Praecipice into which they will surely fall . . . if everything for which the Learned are not able to give satisfactory Account shall be condemned as false and impossible, then the world itself will seem a meer Romance.

Dyer then enunciates the distinction between the supernatural and the paranormal: 'the existence of Spirits cannot be found by Mathematick demonstrations, but we must rely upon Humane reports unless we will make void and annihilate the Histories of all passed things' (ch. 5).

The tragic irony of *Hawksmoor*, however, lies in the fact that while in terms of the novel's methodology Dyer is justified in his belief in a spiritual universe, the world-view which reinforces that belief is as dark and despairing as that in Hogg's *A Justified Sinner*—a book which *Hawksmoor* resembles not only in tone and structure but also in its blend of historic accuracy with artifice. For Ackroyd invents a seventh church for Dyer/Hawksmoor, Little St Hugh's in Moorfields: the dedication to the medieval choirboy, whose murder led to a persecution of the Jews, is cruelly appropriate. The manipulation of history in the name of an invented meta-history makes this novel a very overt piece of fiction; yet paradoxically it induces the feeling that it is a corrective to history. Whatever else, it is a dark vindication of the truthfulness of imaginative vision, offering no resting-place for those in search of absolute truth, only a sense of moral involvement in infinite relativities that can transcend even the extensions of space in time. Is not church architecture a systematized imaginative construct embodied in forms that set out the inner workings of a spiritual universe? What Dyer posits of his temples dedicated to the power of the supernatural can apply no less to those at Canterbury or Chartres.

The continuance of long-established supernaturalist traditions into the present time indicates the duality of twentieth-century responses to a subject which previous ages would have looked at very differently. The freedoms of naturalism which have opened up every aspect of the material world to fictive presentation in the face of what seems like every residual moral or emotional taboo only makes more plain the existence of two polarized responses to the claims of spiritual experience. A closed materialistic outlook keeps such experience under lock and key: since matter is regarded as the primary reality, spirit, in whatever manifestation, can only make itself known in intrusive or demonic forms: it is some such point of view that motivates all negative tales of the supernatural, such as *The Green Man* or *The Woman in Black*, just as it does in the work of Le Fanu or Henry James. But there is an older way of regarding and experiencing the world, and this is the magical imaginative view represented in the Neoplatonist, the alchemical, and the sacramental Catholic traditions, for which matter is itself the expression of spirit, finally only to be known (even if never to be finally understood) in corresponding terms: the physical world is both an image of the invisible one and an aspect of it, just as the various religions and poetic myths may be deciphered as the various dialects and grammars of a common language which constitutes the human response to the unknown in which, in both senses of the term, it finds itself. Such is the outlook we find in writers like Neil Gunn and Alan Garner, expressed in positive terms through a fictive structuring in which such a philosophy may seem appropriate. But the worlds they describe are those of primitive or bygone ways of life.

Hawksmoor, if less prosaic, is in some respects a more genuine reflection of its time. In its imaginative urgency and technical assurance Ackroyd's novel is a powerful embodiment of the perplexity, the embarrassment, and the unease with which late twentieth-century Western writers confront any intuition of the mysterium. Not one of those authors who have devoted themselves to an imaginative investigation and expression of supernaturalist ideas has been able to realize more than the essential ambiguity of this aspect of human experience. Each of them has found the mysterium to be a mirror, a reflection of their worlds and selves through which they pass into the Red King's country, unsure as to whether they are to find themselves the dreamers or the dream.

10

THE EARTHING OF THE SUPERNATURAL

> Simultaneous composition on several planes at once is
> the law of artistic creation, and wherein, in fact, lies its
> difficulty.
>
> (Simone Weil, *The Need for Roots*)

Encountering the Supernatural

'What is before me, I see.' Miss Bates's gentle boast in *Emma* is
in her case fully justified: certainly no other inhabitant of Highbury
gives such correct interpretations of what goes on before their
eyes. But if they did, this particular book would be a poorer
thing. Even a naturalistic novelist like Jane Austen needs to explore
what goes on behind appearances, searching out unseen motivation and allowing for the hinterland of dream, reverie, impulse,
and unprompted memory that Virginia Woolf describes as
'the under-mind'.[1] This is the world of intangibles (and thus
immeasurables) in which human beings pass both their waking
and their sleeping hours, even when it is the rationalized world
of conscious motivated activity which, for practical purposes,
they designate the 'real' one.

Novels that concentrate on the world of the under-mind portray human society as an interaction of subjectivities. These
subjectivities, while agreed as to the validity of sense impressions, also partake of a shared world of intangibles, which are
assembled, deciphered, and transmitted in narratorial form by
the responsive and discerning imagination. Attuned to external
appearance through metaphor and symbol, this patterning is at

the heart of all imaginative writing; and supernaturalist fiction is a recognition and an exploration of that centrality. Far from being an eccentric species of novel, this one is a dramatizing distillation of what every novel by its very nature has to be.

The supernaturalist novel is, however, eccentric to the extent that it is relatively scarce. The Gothic novelists manipulated one particular literary convention; and to this day they have had their imitators and their own specializing readership. But of the twentieth-century novelists described above, only Machen, Blackwood, Lindsay, and Williams confine themselves exclusively to supernaturalist themes. And among the other writers one finds the supernatural restricted to shorter fiction, as in the cases of Le Fanu, Stevenson, James, and Bowen; or else confined to one or two novels in an author's total output. However, the concept of the supernatural can animate a novel as well as control it; and in works as different as *A Glastonbury Romance*, *A World of Love*, and *The Lion of Cooling Bay* we find the same kind of pervasive presence infiltrating and interpreting events as we do in *Wuthering Heights* and, less ostentatiously, in *Jane Eyre*, *Darkness Visible*, and *Hawksmoor*. In each case one is made conscious of an encompassing and providential mystery.

This kind of novel displays tensions between the states of irrational feeling and the material level of reality which are analogous to those between supernaturalist and naturalistic fiction as a whole. The former species portrays, however imprecisely, a metaphysical extension of that world of collective representations on which both traditions draw; but fictive naturalism confines itself to the world of rationalized behaviour and predictable social laws. However, by exercising the powers of the imagination while at the same time tacitly acknowledging that the universe is mechanistic, such a project is inherently self-contradictory. A gap appears to open up between the aspirations of the measurer and what is presumed to be the nature of the measured; and it is this gap which the finest supernaturalist fiction seeks to bridge, to explore, or to eliminate.

Crowding through it come the various spectres, elementals, and sensory dislocations which are these particular fictions' raw material. Figurative expressions of a lost awareness of the primacy of spirit, they were formerly natural metaphors but have now become outlandish similes, as though a friendly fireside Lob were to transform itself into a poltergeist. The reifying of the spiritual

order, which inevitably follows from the mechano-morphic processes of materialist philosophy, results in a conception of spirit
as a quasi-material force with localized embodiments in a multi-
tude of guises. To identify the immeasurable with a particular
form, to equate the absolute with a particular ideology, is an
aberration that, because of the loss of belief in the originating
primacy of spirit, is now accounted normative. But it is precisely
such limiting collocations which constitute idolatry, motivated in
part by what Iris Murdoch refers to as the 'authoritarian aspiration to a unique, systematic truth'.[2] It is against such ambition
that imaginative fictions make their implicit protest.

The paradox is that spirit should nevertheless depend upon
material embodiment if it is to be communicated as spirit. In
supernaturalist literature, the more integrated with physical appearances these embodiments are, the more disquieting and
memorable. The writers of ghost stories in particular are skilled
in producing disorientating effects, such as a clock still ticking in
an empty locked-up house; or the swish and crackle of a brush
upon an unseen head of hair; or a sound 'more close to us than
any separate being could be'.[3] However, these disturbances to
physical and nervous expectation are but so many means to an
end; the more ambitious supernaturalist writers, the novelists,
while using a wide variety of imagery and literary motifs, aim at
effecting nothing less than a total reorientation of epistemological priorities. John Cowper Powys is quite open about his intentions in this respect; and Machen, de la Mare, Bowen, Williams,
Le Fanu, Reid, and Murdoch are similarly concerned to discour-
age any tendency to regard bodily and mental modes of consciousness as separate and distinct from one another. Whether
by frightening their readers or by outraging their preconceptions,
their fictions demonstrate through instruction, lyrical invocation,
or inquiry, that it is through a divided awareness that spiritual
forces come into play, for evil or for good.

Generally speaking, these novelists' depiction of such intrusions, eruptions, or transfigurations reveals a threefold process at
work. In the first place they portray an experience of *visitation*,
the impact of the supernatural as invasive, alien, challenging,
external, hostile, or subversive: its peremptory character acts
as a bridge thrown over from one mode of consciousness to
the other. A materialistic viewpoint necessarily experiences the

supernatural in some such way, and is thereby faced with a challenge to its presuppositions. At its crudest, this process occurs in the ghost story, as when Marley appears to Scrooge (that embattled sceptic); but visitation can take on more sophisticated forms, as in Kipling's poignant 'The Gardener'. It can also be terrifying: in Charles Williams's *The Place of the Lion*, superhuman forces demoralize and devour those vulnerable to them through their moral and spiritual inadequacies; an arrogant young literary scholar, for example, is confronted with the image of her sterile, self-referential attitude to learning in the form of a huge stinking pterodactyl.

In another guise, visitation can arouse the desire for knowledge, or a responsive worship, the former involving esoteric occult studies, as in the novels of Bulwer-Lytton, the latter an intrusion of didacticism, as in those of C. S. Lewis. Or the visitation can occur through the impact of physical environment, involving the revelation in material form of a spiritual reality that transcends its own medium. (Many such epiphanies are to be found in non-supernaturalist writers as well, 'nature' being a popular repository for other-worldly emotions.) In psychological experience, however, visitation can be interpreted, wrongly or not, as hallucination. The enigmatic happenings and appearances in the novels of Phyllis Paul, for example, deflect the reader's psychic bearings through a form of veridical obliquity, while confronting the fictional characters with their own moral obtuseness and powers of self-deception. As to later twentieth-century novelists, while they treat the solicitations of the supernatural as material for speculative discourse or verbal play, in their fictions no less than in the others, visitation elicits a response, in part conditioned by historical and social circumstance, in part by personal temperament and choice, a response which provides each novel with its particular subject or story. Its object is to awaken readers out of imaginative and mental sleep: but that awakening and resultant insight is not its subject's peculiar property, nor is to be taken for an absolute.

The supernatural is also portrayed as the experience of *possession*, the absorption of the under-mind by forces productive of ecstasy, spiritual illumination, dreams of power. This enables the supernaturalist writer to explore in depth the nature of the gap in consciousness. Here too the several literary methodologies exhibit a variety of exemplars. Often it is the haunted as much as the haunters who are disturbing, most especially in the work of Walter de la Mare; while the demonic agencies that wreak such havoc in the tales of Le Fanu and M. R. James emerge from a without that in some cases seems alarmingly like a within. In the hermetic tradition also, possession can be dangerous, the spiritual energies overwhelming those who dabble in powers beyond their control: *hubris* destroys both Lytton's Margrave and Blackwood's Philip Skale, and there are watchers on a diversity of thresholds, witness the fate of Meade Falkner's John Maltravers. In the theological tradition the internalizing of the supernatural is expressed through psycho-drama, as in the finest work of George MacDonald and Charles Williams, wherein the soul is judged and chastened in order to be possessed by God. (Or, as in the case of Wentworth in *Descent into Hell*, to be damned through its choice of self-possession.)

Physical environment likewise can be internalized through reverie and mythologically prompted contemplation, as when John Cowper Powys, for example, writes of 'a lonely tarn, that might be called Rhiannon's Tarn, with ice-cold water & dark tussocks & old dead bull-rush stalks, far away did it carry my mind, far far away'.[4] This is a haunting and not uncommon experience, such as one finds recorded also in the pages of Neil Gunn and Forrest Reid; yet it is no more frequent than are the subjective demonic presences delineated by Henry James (with such scrupulous deliberation) as negative aspects of the implacability of conscience, or by William Golding and Iris Murdoch as unexplained upheavals in customary mental processes which can give rise to moral conflict and decision. In these cases the undergoing of possession is an experience of potential danger, an exposure to truth and a testing by it. At the same time the experience in whatever form is not itself an absolute. These writers' whole object is to transcend the self-imposed limitations of humanistic naturalism.

Of course such patterning and allocations are to some extent arbitary and are always oversimplified. The presence of the supernatural is more completely evident in the *interaction* between visitation and possession. Whereas visitation may be viewed as the work of chance, whether lucky or malign in its random nature, the experience of possession suggests predestination, moral helplessness, inexorable fate. Neither interpretation by itself encourages freedom of action or personal responsibility. These things are the result of recognizing that the two ways of interpreting the supernatural are subordinate to an acknowledgement of its general providential, ultimately meaningful presence. Much of the literature that dramatizes this process of interaction, between externalized projection and subjective feeling, keeps these two elements in tension and thus allows for and suggests a supernatural state transcending both: *Wuthering Heights* affords a good instance of this process. The recognition that ultimate truth transcends any systematic definition allows the reader to respond in freedom from the author's conscious or unconscious supervising presence. The capacity to scrutinize the invasive phenomena while at the same time responding to them from within the confronted psyche leads to the recognition that the supernatural is precisely that—super-natural—a recognition that affirms and refines and fulfils the more restricted viewpoint of materialism. Such an insight alone protects its recipients from the worship of false gods in the shape of pre-emptive literary or sociological theory.

All the most persuasive accounts of the operations of the supernatural are characterized by this interaction between visitation and possession, taking place in the context of realistically recorded human experience and behaviour. The process occurs in works as diverse as *The Return* and *Nuns and Soldiers*, *Dracula* and *The Hill of Dreams*, *The Turn of the Screw* and *The Place of the Lion*, *Twice Lost* and *A Glastonbury Romance*. In each case the visitation confronts its recipients dramatically, even while emerging from their inmost selves. The encounter is experienced as preternatural, yet belongs to the paranormal realm, that spiritual universe of archetypes, memories, and dreams that encompasses and conditions human knowledge and responses, in which human beings are at home and pass their time, and in which they are attuned in their varying modes to the presence of the mysterium.

The Refutation of Single Vision

If the supernaturalist element in a novel be interpeted as an intensification and a particularizing of this under-mind or inward consciousness, then its literary strategies serve as questionings of what William Blake denounces as the 'single vision'[5] of materialistic popular responses, and as a reordering of it. Indeed, the very history of supernaturalist fiction charts the exposure of nineteenth- and twentieth-century rationalist philosophy to the concept of the unseen; with equal validity one might interpret it as the continued assertion of a previous, less mechanistic reading of human experience, as against the disjunction of physical from spiritual consciousness implicit in a technologically motivated response to life. And what happens in the literature reflects what happens in contemporary awareness: where the development of the English novel as a literary form is concerned, here too the growth and adaptation of supernaturalist themes provide the plot with a meta-text or story.

Subjective consciousness was not regarded as of epistemological relevance in the eighteenth-century Enlightenment, with its concern to maintain a civilized, increasingly humanitarian society, and with its stress on right reason, good manners, and judicious piety; in *Northanger Abbey*, Henry Tilney's rebuke to Catherine Morland for her vain terrors and their consequences springs from that belief in common sense and material progress which underlies the ideals of the kind of naturalistic fiction that Jane Austen wrote.

What have you been judging from? Remember the country and the age in which we live. Remember that we are English, that we are Christians. Consult your understanding, your own sense of the probable, your own observations of what is passing round you—Does our education prepare us for such atrocities? (ch. 9)

Were it to be put in the twentieth century the question might not be so rhetorical; but it is in any case quietly modified at the novel's close when Catherine, having learnt from Henry of his father's greed, credulity, and petty-mindedness, had 'heard enough to feel, that in suspecting General Tilney of either murdering or shutting up his wife, she had scarcely sinned against his character, or magnified his cruelty'. The reversed irony is an instance of the author's grounding in the fiction of her predecessors; but it also suggests a renewed consciousness of the inseparability of physical and metaphysical experience.

That inseparability is a central affirmation of Romanticism, in which the supernaturalist novel was to serve the function of questioning both the precise power and status of right reason, and its relevance to the expanding conception of what was possible to human psychic and emotional capacities. But this romantic conception of nature as beneficent and liberating was to be marginalized and rendered trivial in the shape of random individated ghosts from the past (the sentimentalizing of folklore, the cult of the Gothic picturesque, the craze for occult studies, seances, and so on), all subordinated to the 'common sense' attitudes of popular utilitarian philosophy; the early tales of Arthur Machen and E. M. Forster, for instance, reflect on just such a process towards the dismantling of belief in a supernatural world and its all-encompassing significance.

Where moral evaluation was concerned, the division between physical and spiritual categories widened still further in the Victorian age, despite the protests of social observers and aesthetic theorists such as Ruskin and John Stuart Mill. The religious tendency to postpone the blessings of the spiritual world to a hypothetical 'next' one (that Promised Land to which Evangelical piety looked so wistfully forward) led to a kind of emotional schizophrenia. An industrialized and commercially expansionist society found itself with an essentially materialistic spirituality, whose careful if often plaintive separating of this world from its metaphysical roots, when coupled with sexual self-repression and social and economic guilt, caused discarnate spirituality to turn demonic in the pages of the later nineteenth- and early twentieth-century preternaturalist writers. The proliferation of literary ghost stories at this time reflected the fragmentation of a wider, deeper spiritual consciousness.

For their part, the huge cloudy intellectual edifices of occult studies and the more precise and self-confessedly provisional absolutes of Catholic theology expanded into (or were diluted with) a heightened consciousness of the potentially supernatural character of matter in itself—a cardinal precept in the thinking of G. K. Chesterton and Charles Williams; but the religious awe experienced by many Victorian scientists disseminated as a pseudo-paganism. When religious traditions are assumed into a reawakened aspiration towards a communion with natural energies and planetary laws, bodily instincts come to be seen as surrogates for inner spiritual experience, as tends to happen in the later writings of D. H. Lawrence. (But in this respect, as in so many others, Lawrence is not to be nailed down. He had an intuitive understanding of transcendence, witness the ending of *St Mawr*.)

An interiorized equivalent of this process may be observed when the psychologizing of the creatures of mythical imagination and poetic lore absorbs the supernatural into the realm of subjective experience, leaving its symbols and rituals to become the sport of an intelligence rational enough at least to acknowledge such rationality's limitations: in the work of Golding, for example, one sees the supernatural subordinated to the novelist's manipulative concern with intellectual discourse. But, as Jung pertinently remarks, 'Statements concerning possibility or impossibility are valid only in specialised fields; outside those fields they are arrogant presumptions.'[6] The position of the would-be refuter of any concept of the supernatural has come to resemble that of the scholar who claimed that the *Iliad* was not the work of Homer but of another poet with the same name.

In the later twentieth century the loss of any generally accepted theological or metaphysical infrastructure led to a revived stress on inventiveness, on the self-consciousness and autonomy of fictive processes as such; and this insistence on the self-propagating linguistic basis for imaginative projects, perhaps because of that very loss, points to the existence of the supernatural through the implications of its absence. In this respect the history of the supernaturalist novel provides a metaphor for the twentieth century's ambiguous attitude to the concept of objective truth.

Whether incidentally, as in the work of the greater novelists, or in the specifically supernaturalist ones as an element in the corpus of fiction as a whole, the presence of a supernaturalist theme serves to check the presumptions of materialistic naturalism. For no novel that contains an element of the strange and unaccountable can readily be confused with absolute replication: it is avowedly fictive, and distances the responsive assessing intelligence by its self-assertion as an artefact. It is easier to idolize an emotion or a concept than to idolize an object; and just as the carved images in a Catholic church are blatantly images and most clearly not the reality they signify, so the supernaturalist element in a work of fiction safeguards the distinction between the signifier and the signified. It is to that extent an instrument of deconstruction. By making ostentatious the fact that a novel is no less an artificial construct than is a poem, it reminds us that all language is metaphorical, and that the truth is never to be captured in a verbal formula, but is only to be intuited in the space between the signifiers.

Thus the status of supernaturalist fiction within the novel as a genre mirrors the condition of the novel itself when regarded as a veridical and verifiable means of fictive statement. Indeed, the ideal of an absolutely true record of events is unattainable in an objective sense, since the recollection and ordering of every narrative is subject to the mental and emotional limitations of the narrator. As George Steiner comments, 'There is a sense in which no human discourse, however analytic, can make final sense of sense itself.'[7] But if there can be no absolute truth discoverable by human beings and thus portrayed by them, there may be an absolute truthfulness in their several attempts to enunciate it. A novel is no more and no less fictional in essence than any other record of human behaviour. What tends to confuse the issue is the literal-minded reader's fallacious insistence that a fully objective record may be possible. Such insistence when applied in practice leads to the worship of false gods, and can only result in muddle, a muddle compounded by the booming emptiness which is all that such confused response can make of the enveloping source of human life that we experience as mystery.

The Relativity of Absolutes

It is because supernaturalism is a rogue element in the house of fiction that it can function as a shatterer of idols. On the one hand it assaults the tendency to confine the reading of experience within any one given philosophy or framework; it qualifies established orthodoxies; it can be both critical and corrective. At the same time it asserts, and at its finest it demonstrates, an enlargement of possibilities, opening up epistemological and imaginative horizons, extending the measurements perceptible to narrowly rationalistic philosophy. It is of course wide open to scepticism, to mockery, and to deconstruction; but scarcely to refutation. Whether derided or taken seriously, supernaturalist fiction permits a variety of interpretations; and whether presenting its material as visitation or as possession it can be read both as hermetically decipherable and epiphanic, or as judgementally invasive. Neither category, however, can be regarded as definitive.

Where the correction of customary perspectives is concerned, in turning to the tale of terror, for example, one can see how it aims to destabilize social as well as personal assumptions, to invite a reordering of expectations and to alert one to the possibility of alternatives to the routine adherence to regulated custom and clock time. It was to persist as an undercurrent of unease as the myth of conjoined material and ethical progress propelled the nineteenth and early twentieth centuries towards the First World War; since when its corrective energies have dispersed in the forms of science fiction of the critical and prophetic variety, or in those novelists of inner questioning and disturbance who concentrate on the individual consciousness. The secularized humanism that sustained the high-thinking ideals of the cultivated middle and upper classes is questioned from within by a literary tradition no longer content merely to reflect or satirize. The moral soundings of even so unpretentious a novelist as L. P. Hartley reverberate beyond their formal occasions: they alert one to the haunted imaginative landscape in which the twentieth-century English consciousness has its being. The more robust, because more externalized, negations found in supernaturalist writing since 1945 represent in their turn an attempt to clear the air of the receding nineteenth-century intellectual disciplines and emotional constrictions: they break down the unreal barriers between 'exterior' and 'interior' experience. In this purified and winnowing atmosphere post-modernist writers have exercised their imaginations unfettered by previous formal sanctions and, on occasion, even previous moral ones.

In purely literary terms, affirmative supernaturalism has been less influential than its corrective counterpart. The alternatives to materialism offered by the outlook of hermetic or theological novelists have found a more restricted readership, neither tradition by its very nature having been able to command a major, because ideologically uncompromised, fictive talent. Moreover, affirmation tends towards the building of defensive mental structures; and the most persuasive supernaturalist writers of this kind are those who, like Kipling, are impelled by a spirit of curiosity. Kipling indeed is a classic instance of an author who can purvey a sense of mystery without any doctrinal attachments.

What all affirmative supernaturalists do induce is a sense of natural piety: it is a persistent undercurrent, for example, in work as different as that of George MacDonald and Forrest Reid. But this basic quality of a humane civilization has been a victim of the break-up of familial, religious, and social structures, and as the twentieth century proceeded, piety was to commend itself more generally in terms of place, of landscape, and of physical

and historical rootedness; one finds it lending depth and resonance to writer after writer from the time of Hardy on. In this
respect John Cowper Powys is a significant figure, since his handling of traditional supernaturalistic material combines a scepticism concerning previous interpretations with a committed belief
in the existence of a transcendental world of the imagination
which is rooted in the life of everyday people and affairs, and is
thus available for all to enter.

It is the pre-emptive scepticism, however, which makes that
belief in a transcendent world persuasive: as against the other
affirmative correctives, the celebration of the physically numinous
does not spring from, nor direct one towards, a closed metaphysical or moral system. And indeed, where supernaturalist
writers are in question, while each methodology undermines the
closed world of materialistic theory, each one in turn tends to
qualify the others: they thus confront one with the relativity of all
pretensions to a definitive access to the truth.

Thus on the one hand while the tale of terror and the hermetic
and theological traditions deliberately convey a frontal challenge
to established certainties, the last two also propound counter-
certainties, depicting the coinherence of body and spirit in a
declaratory manner that concedes the terms for its own rebuttal:
they exhibit in matters of the spirit the vulnerability of any such
ostensible objectivity to rationalistic refutation. This is a literature too comfortably subordinate to its own initial premises.
But all three literary traditions lay themselves open to rejection
on materialistic grounds.

On the other hand, the novelists who explore the mystique of
earth-life aim to close the division between materialistic and
ideological philosophies through the medium of subjective feeling, a less readily impeachable basis for belief; yet even here
one finds Lawrence voicing a qualified despair with regard to
the accessibility of any ultimate truth, and Powys concluding *A
Glastonbury Romance* with the stoically non-committal words
'Never or Always'. Other novelists, such as James and Bowen,
internalize their material psychologically, inducing acceptance of
the supernatural through an imaginative sleight of hand and a
tendentious stylistic rhetoric; but such conjurations are in turn
exposed and exploited in the work of late twentieth-century
novelists—Golding and Spark, for instance, effecting a demolition
of idols and the reuniting of writer and reader in an arbitrary

game of 'let's pretend'. They approach the mysterium obliquely, by indicating its presence through a deliberate trivializing of its absence. But these various contradictions, counter-assertions, and refutations are themselves all relative. There is no guaranteed way of verifying or assessing the conclusions which these several writers individually attain.

Naturalism Transfigured

The literary relationship between natural and supernatural is, however, reciprocal. Again and again touches of realism guarantee the authenticity of the extraordinary, as Horace Walpole had foreseen that they would: indeed, the quotidian aspect of fictional narrative heightens the extra-temporal one, and the tension between them enlivens the novels in which they are at odds. A supernaturalist story needs a naturalistic setting, for as Woolf remarks, 'Some degree of reality is necessary in order to produce fear, and reality is best conveyed in prose.'[8] When the supernaturalist material preponderates over the everyday the results are imaginatively deadening. The struggle between Victor Frankenstein and his creation mirrors the tension between the dictates of naturalism and the demands of natural forces and what human responses have made of them; between the dictates of reason and the overwhelming demands of the subconscious; between the separateness of arbitary fantasy and its integration into a world of responsibility and human claims. These tensions safeguard fictive representations from the limiting extremes of naturalistic materialism on the one hand and the suffocating solipsism of interiorized narration on the other.

It has to be admitted, however, that the masters of the supernaturalist genre who specialize in this borderline between the subjective and objective—Le Fanu, Machen, de la Mare, Williams —because of that very concentration of concern are necessarily restricted in their appeal. They are writers who mean a great deal to a relative few, but whose compass does not extend very far beyond their particular inner worlds. Yet it is also because of that concentration that they are so authoritative in the field they occupy. Like the literary tradition to which they belong, they act as jokers in the pack of materialistic emotional and fictive expectations; and it remains true that those greater novelists whose imaginations, while remaining totally attuned to the existence and character of the material world, likewise encompass the realm of spiritual experience, are the more rewarding for that particular inclusiveness. Scott, Dickens, Hardy, Kipling, Conrad, Lawrence, would not be the major writers they are without the apprehension of the mysterium which haunts their finest work. It is this involvement which universalizes them, pungent and thus potentially alienating though their stylistic idiosyncrasies may in many cases be. Their worlds are not closed worlds as are those of lesser novelists.

For if physical experience is the basic material of fiction it is characteristic of the most satisfying English novelists that they describe and celebrate the visible and tactile so graphically that they can engender a feeling of enhanced significance in familiar subject-matter, and to that degree exhibit the presence of the mysterium latent in material things and in the ordinary exchanges and individual experiences of human life. It is what all the major writers in these various traditional approaches and literary methodologies have in common—the power in their several degrees to effect an imaginative transfiguration.

The immanent energy that constitutes the visionary power is evoked and dramatized explicitly in supernaturalist fiction, where it is set in a dynamic relationship with 'single vision', the philosophy that would confine the art of the novel to the physical and mental realms alone. The whole thrust and achievement of supernaturalist fiction is thus a potential vindication of the romantic viewpoint against the deadening constrictions and self-destructiveness of an exclusive rationality. However much the supernaturalist writers may emphasize the unfamiliar and the uncanny, the finest of them do so in order to heighten and enrich their readers' responses to the familiar and the seen. They enhance those readers' susceptibilities to the endless diversity of life in space and time: by venturing into the realm of the invisible they renew appreciation of the significance of the life of the five senses.

The menacing or paradisal, but always substantial, landscapes of Ann Radcliffe and John Cowper Powys share in this transforming energy no less than do the densities and crooked indirections of Le Fanu and Peter Ackroyd, or the rustling inconclusiveness of Walter de la Mare. It is the same energy that lies behind the restive curiosity of Rudyard Kipling and Iris Murdoch, and sets off the psychic alarm bells of Arthur Machen and Montague Rhodes James. It sounds a note of disquiet, suggesting possibilities that may be liberating or may be subversive, but a note which in either case can shatter the structure of the closed materialistic universe. That shattering is for a purpose. Its aim is to accommodate our perception of the world to the ways and conditions under which we actually perceive it, and thus to make it possible to feel at home there. This may sound a paradoxical intention; yet it is this ambiguous perspective upon physical reality which is the distinguishing feature of the English fictive imagination. It is certainly appropriate, and perhaps inevitable, that it should colour even the naturalistic novels of the island which Roman legionaries declared to be inhabited by ghosts.

EPILOGUE

AN ABOLISHING OF IDOLS

in absence you receive the essence

(Mary Casey, 'Tao', *The Clear Shadow*)

The query with which this book began has finally to be confronted. Is the supernatural an authentic part of human experience, to be taken seriously; or is it a delusion? Put baldly like that the question can only be confidently answered by the religious believer or the equally assured atheist: it is a matter of personal conviction rather than one susceptible to exclusively rational inquiry. Where literature is concerned, therefore, rather than address the question head-on, it will be more to the point to ask whether the supernatural is a meaningful category in our contemporary critical discourse; and if it is not, does this then constitute an enhancement or a restriction of our capacities for interpretation and response? Only when those questions are met can one begin to determine to what extent the concept of the supernatural corresponds to a reality that we can know.

Such questions can have no appropriate answer without an acknowledgement of their metaphysical implications; but few people at the present time are prepared to take metaphysics seriously or to see in them more than an arid, outworn irrelevance. At best they are regarded as a form of intellectual sport. However, this stress on play is nothing new: it is no less an aspect of traditional religious understanding than it is of twentieth-century linguistic relativities. Mystics, saints, and theologians have always been well aware of how provisional all verbal formulations of metaphysical reality must inevitably be, even while conceding that it is the function of poets and storytellers through verbal means such as parable, imagery, and song, to keep alive that feeling of life's illimitable mystery which is metaphysics' energizing source.

Where a novel's readership is concerned, one notes a variety of response. From its earliest days the evidential value even of naturalistic fiction has been called in question. Anne Brontë herself is not absolute in her claim for the instructiveness of supposedly true stories, admitting that 'in some the treasure may be hard to find, and when found, so trivial in quantity that the dry, shrivelled kernel scarcely compensates for the trouble of cracking the nut'.[1] Such a metaphor implies a rigorous view of novel-reading, one that asks for more justification for the practice than nowadays most people outside of the academies would require.

For example, the majority of readers probably turn to fiction in order to be diverted: they ask not so much for instruction or even for plausibility as for pleasure. Such pleasure is at once atavistic and sophisticated. The former attribute explains why E. M. Forster's self-styled 'droopy regretful' admission that 'Yes—oh dear yes—the novel tells a story'[2] should perplex readers unconcerned with literary theory or the practicalities of authorship. They can see no reason why such a statement need be in any way apologetic. They are perfectly well aware that what they are reading is 'only a story'; to that extent they are undeceived sophisticates. Thus being entirely credulous with one part of the mind while entirely sceptical with another, the undemanding ghost-story addict, however irreligious in outlook and materialistic in imagination, can still register a pleasant shudder at the thought of the weary travelling salesman who, slipping gratefully into a strange bed, feels his leg being clasped by a cold and slimy hand. Such fancies, it might be argued, are the closest that some materialists may get to a feeling of religious dread. Their only experience of the supernatural is of its being an abnormality, and as readers they welcome it, as providing a bit of spice to ordinary life.

Other readers are attracted to prose fiction because of its representative quality, its historical value, and its reflection of social conditions. For them it calls for no defence, since they are not reading for distraction but for instruction; they, presumably, are the readers Anne Brontë was seeking to placate. Rather than tease out the distinction between the factual and the imaginary, they ignore it. For such readers, novels and tales about the supernatural have little appeal save as psychological or social curiosities;

while as critics they will be less concerned with the subject's persuasiveness than with dismantling its pretensions and placing it in the context of linguistic conditioning or social expectation. In their case the supernatural is paranormal merely, and the question of separate ontological status does not arise.

Readers more attuned to what creative writers believe themselves to be engaged upon appreciate fiction primarily for its artistry; they base their evaluations on what they judge to be a novel's moral or imaginative truth, its enlightening quality and power to instruct through extending the capacity for responsive and discriminating perception. For them, all fictional material, whether supernaturalist or not, calls for serious attention: their reading is a reading of the material, a deciphering of what is implicitly the 'mystery' of this particular craft. At the same time their attitude to the truthfulness of fiction in all its aspects is both positive and negative. It is positive to the degree that it encourages the fullest recognition and response, yet negative in its insistence that such an absolute response is itself provisional. Their reading is disinterested and not designed to confirm or to ferret out particular attitudes or their own beliefs.

Whatever the degrees of responsiveness granted by their readers, the writers themselves assume a possibility of valid communication. Airy talk about 'language games', while fully justified theologically in relation to the ultimate unknowable absolute, is, in an atheistic or irreligious realm of discourse, merely self-cancelling: there is a world of difference between an acknowledgement of epistomological limitation and a carefree propagation of futility. Indeed, trust in the validity of language as a mediator of a truth that transcends the mere desire for communication applies to any critical or philosophical contention of a generalizing kind, however negative its purport. Accordingly, a novelist's engagement with the mysterium is an act of faith, not only in the possibility of communication but also in the experience to be communicated, even allowing for the final unknowability in conceptual terms of the source of that experience. In this respect fiction is to life what life, when theologically considered, is to the metaphysical dimension that transcends it: it may replicate it or point towards it but it cannot be equated with it.

The various ways in which the supernatural is portrayed in fiction are themselves conditioned by the various literary forms through which those accounts are mediated—forms that in their turn are the products of the social preoccupations, publishing conditions, and literary fashions of their time.[3] All this under-lines the fact that it is misleading to try to separate the exam-ination of supernatural phenomena as such from the psychic, social, and economic conditioning under which they manifest themselves. To look for supernatural reality apart from material reality, or (no less misleadingly) to equate it with that reality, is to resort to extremes of credulity or scepticism, both of which in their contrasting ways constitute idolatry, through the taking of the part for the whole, and the relative for the absolute. The authenticity of the supernatural is not to be established on such simplistic terms as these.

That this is a matter of contemporary significance is evident from the tendency of late twentieth-century Western societies, when confronted with metaphysical concepts, either to embrace absolute fundamentalisms of all kinds (witness the current pro-liferation of religious sects and the flourishing of materialistic superstition) or to relapse into the sort of frivolous scepticism nurtured by large sections of the popular press and entertain-ment industry. A novel, however, by its very nature testifies to the limitations of such restrictive responses; it is veridically ambigu-ous, being faithful to observed reality and 'true to life' while at the same time being patently fictitious. Within the text itself a similar duality colours the part played by supernaturalist themes, and this is one which, as we have seen, serves to focus the issue of fictive truthfulness in a singularly testing way.

The paradox inherent in the notion of absolute transcendence (the paradox that its being must be predicated while forever re-maining unattainable) extends to areas other than theology or art or deconstructionist philosophy: it is not merely an intellectual concept but a matter of common personal experience. In imag-inative fiction the presentation of the supernatural accordingly emphasizes its effects rather than attempt its definition. Those effects have been discussed in the preceding pages, in which it is maintained that the supernatural is portrayed by these writers as manifesting itself in both supportive and subversive aspects, and as being both immanent in human nature and invasive of it. The finest of these novelists describe the interaction between these two aspects, in the process exhibiting the supernatural not as a mere excrescence upon human life but as its vitalizing heart, and as coextensive with nature though not coterminous with it.

Accordingly the depiction of the supernatural in whatever guise tends to refute the narrow perspectives of rationalistic material- ism, and challenges the excessive cerebration that underlies so much contemporary thinking and critical procedure. By the same token, through its frequent outlandishness it serves as a check upon the embracing of false absolutes, encouraging the recogni- tion that all such personalized apprehensions of what is regarded as the truth are necessarily provisional. The sense of the super- natural conveyed by the work of so many imaginative writers braces the mind against all that is doctrinaire and partisan, every- thing that denies the nature of the mysterium. But it rests on a basis of normality: supernaturalist fiction is to be distinguished from mere fantasy.

One characteristic species of ghost story is that in which a person alone in a familiar room will be looking in a mirror, only to see in it the reflection of a second figure. This figure, the unknown quantity, the embodiment of the irrational, has many fictional embodiments, and is invariably subordinate to some overriding purpose that exists beyond itself and which operates the ensuing action, working towards some alteration or solution of existing states of being.[4] The fact that the supernatural mani- festation is subordinate to this originating source behind it, and is clearly not to be confused with it, is an inbuilt safeguard against the divisive and destructive worship of idols. Indeed, by its osten- tatious dramatizing of spiritual energies and conflicts, the supernaturalist literary tradition may be said effectively to con- tribute to their abolition. For it is the very nature of such stories that their meaning does not inhere within themselves. They implicitly support the contention that what essentially 'is' must be intuited through the defining silence of 'is not'—another in- sight in which theology and deconstructive linguistic philosophy coincide. Despite which, 'Without words there can be no silence.'[5]

The power to suggest a metaphysical dimension behind ma- terial appearances is not confined to specifically supernaturalist writers. The idea that existence is conditioned by an encompass- ing negation is a recurring theme, for example, in the work of Joseph Conrad, a novelist who, more than most, reverts to the experience of cosmic awe. Although he does not engage with preternatural happenings as such, few other prose writers project so impressive a response to the unfathomed recesses of the human soul, coupled with so deep a sense of the tragic predicament of that soul at the mercy of the indifferent universe around it. To Conrad supremely, the dimension of mystery is mirrored in the sea, that ocean where 'either an elemental voice roars defiantly under the sky or else an elemental silence seems to be part of the infinite stillness of the universe'.[6] It is a comprehensive evocation, and one that epitomizes the dual impression of presence and of absence, in which all authentic conceptions of the supernatural originate. Certainly a very positive non-presence confronts every effort to apply materialistic tools to the dismantling of those provisional structures erected as expressions of humanity's response to the mysterium. When Pompey forced his way into the Holy of Holies in the Temple at Jerusalem he found it empty; and the attempt to discredit the validity of belief in a spiritual universe through anatomizing its embodiments is as self-defeating as would be the hope of shouting down the echo in the Marabar caves. The absolute is not to be tracked down like some unknown statue in a sacred grove.

Supernatural reality is known to us both by its presence shining through the material universe and perceived by the eye of the responsive imagination; and by its simultaneous absence, known in a sense of inadequacy and loss. It is an aspect of human experience which can be neither definitively objectified nor definitively refuted, for the supernatural becomes real to us precisely as, while acknowledging its presence as fundamental spiritual reality, we refuse to identify it with any single or exclusive human form or artefact or ideology. Its presence is pervasive, at times concentrated, but never as one presence among others. To this extent it can only be realized in its absence. It may be encountered; it can never be controlled. It may be sensed as goodness is sensed, and as truthfulness and beauty are.

Certainly human love has to accept the paradox that the knowledge of presence and the knowledge of absence can be simultaneous: it is a knowledge both of consolation and of pain. In absence is presence, in presence absence: all absolutes are subject to this qualifying ambiguity. The safeguarding of that necessary and protective dual awareness is not only the imperative for human social and individual stability, but is also the mark of any work of fiction which aspires to convey a persuasive realization of the mysterium. The artistry of those novelists who concern themselves with meta-history and transcendence is vested in a judicious but at the same time agile sense of its own inevitable limitations where such questions concerned.

> Will God dispute over words? No; but man
> must, if words mean anything, stand by words,
> since stand he must; and on earth protest to death
> against what is at the same time a jest in heaven.[7]

Those lines enunciate a paradox that is at the root of both philo-sophical and fictive explorations of mystery and the supernatural.

NOTES

Except when stated otherwise, the place of publication is London.

Chapter One: The Joker in the Pack

1. See Winifred Gerin, *Emily Brontë* (Oxford, 1971), 75–80.
2. *Early Victorian Novelists* (1934), 159.
3. 'Poems of 1912–13', *Satires of Circumstance* (1914).
4. Sonnet XIX, 'Methought I saw my late espoused Saint'.
5. F. R. Leavis and Q. D. Leavis, *Lectures in America* (1969), 107.
6. *Heretics* (1905), 99.
7. Iris Murdoch, *The Philosopher's Pupil* (1983), 571.
8. 'Currer Bell', editor's preface to the new edn. of *Wuthering Heights* (1850).
9. Introduction to *Wuthering Heights* (Routledge English Texts, 1988), 30.
10. Review of *A Free Enquiry into the Nature and Origin of Evil* (*Literary Magazine*, 1757).
11. *Conversations in Ebury Street* (1924), ch. 17.
12. e.g. Kathleen Raine, *Defending Ancient Springs* (Oxford, 1967).
13. Cf. '. . . the metaphysics of presence is shaken with the help of the concept of *sign*. But . . . as soon as one seeks to demonstrate in this way that there is no transcendental or privileged signified and that the domain or play of signification henceforth has no limit, one must reject even the concept and word "sign" itself—which is precisely what cannot be done.' Jacques Derrida, *Writing and Difference*, tr. Alan Bass (1978), 280–2.
14. George Steiner, *Real Presences: Is There Anything In What We Say?* (1989), 74–5.
15. Rudolf Otto, *The Idea of the Holy* (1923), 68.
16. *Margins of Philosophy*, tr. Alan Bass (Chicago, 1982), 6.
17. *The Greater Trumps*, ch. 5.
18. See P. N. Furbank, *E. M. Forster: A Life*, ii (1978), 125.
19. See Aldous Huxley, *The Perennial Philosophy* (1947), which provides a useful introduction to both Eastern and Western traditions of mystical philosophy.
20. For a lucid exposition of Christian esoteric symbolism as found in mystical theology and pictorial art, see Lois Lang-Sims, *The Christian Mystery* (1980).
21. W. R. Inge, *Christian Mysticism* (1899), 253.

22. Quoted ibid. 24.

23. 'Taken in the religious sense, that which is "mysterious" is—to give it perhaps the most striking expression—the "wholly other" . . . that which is quite beyond the sphere of the usual, the intelligible and the familiar, which therefore falls quite outside the limits of the "canny" and is contrasted with it, filling the mind with blank wonder and astonishment.' Otto, *Idea of the Holy*, 26.

24. *Of the Laws of Ecclesiastical Polity* (1593), I. xi. 3.

25. *Supernatural Horror in Literature* (New York, 1973), 14.

26. *The Art of the Novel: Critical Prefaces*, with an Introduction by Richard P. Blackmur (1947), 31.

Chapter Two: An Iconography of Fear

1. 'On Common-place Critics', *The Round Table* (1817).

2. In an anonymous essay 'On the Supernatural in Poetry'—*New Monthly Magazine and Literary Journal*, 7 (1826)—Radcliffe also distinguishes between terror and horror, which 'are so far opposite, that the first expands the soul and awakens the faculties to a high degree of life; the other contracts, freezes and nearly annihilates them'. The distinction is a useful one when making critical discriminations between a novel like *Melmoth the Wanderer* and, say, *The Monk*.

3. 'The Spirit of the Public Journals' (1797), quoted in K. K. Mehotra, *Horace Walpole and the English Novel* (Oxford, 1934), 27.

4. 'Walpole', *The Lives of the Novelists* (1821–4) (The World's Classics edn., pp. 188–9).

5. Stephen Prickett, *Victorian Fantasy* (Hassocks, 1979), 59.

6. The majority of Dickens's ghost stories will be found in *Christmas Stories* (1850–67), contributed to *Household Words* and *All the Year Round*. See also *Sketches by Boz* (1835, 1836) and *The Uncommercial Traveller* (1861).

7. *Great Expectations*, ch. 8. The fact that the vision does not come true only serves to make it the more disquieting. The spiritual universe is not here confined by any literary foreclosure.

8. *Selected Essays* (1932), 467.

9. The story first appeared anonymously in 'All the Year Round' 1870–1, and was reprinted as part of 'A Strange Adventure in the life of Miss Laura Mildmay', *Chronicles of Golden Friars*, i (1871), and subsequently by M. R. James in his annotated collection of Le Fanu's shorter pieces, *Madam Crowl's Ghost and other Tales of Mystery* (1923).

10. 'Laura Silver Bell', *Belgravia Annual* (1872). Repr. in *J. S. Le Fanu: Ghost Stories and Mysteries*, selected and ed. E. F. Bleiler (New York, 1975).

11. *Chronicles of Golden Friars*, ii, repr. in *Best Ghost Stories of J. S. Le Fanu*, ed. and with an introduction by E. F. Bleiler (New York, 1964).

12. Ibid.

13. Ibid.

14. In a neat mixture of imperialistic racialism with sexual prurience, it is white women which the participants in the rites of Isis are said to prefer.
15. Michael Cox, Introduction to *Casting the Runes and other Ghost Stories* (The World's Classics, 1987), pp. xxi–xxii.
16. *Supernatural Horror in Literature* (New York, 1973), 102.
17. *Ghost Stories of an Antiquary* (1904).
18. Ibid.
19. *A Thin Ghost and Others* (1919).
20. Cox, *Casting the Runes*, p. xviii.
21. A felicitous account of James's fictive technique occurs in Victor Sage, *Horror Fiction in the Protestant Tradition* (1988): 'his readers . . . are only allowed to follow footprints, the clues, and listen to the accounts of participants. The images are always seen by someone else "out of the tail of the eye"—from the mortal side as it were. The imprint of the spiritual through the material is like the momentary bulge of a face in random folds of cloth' (p. 67).
22. *Ghosts and Scholars: Ghost Stories in the tradition of M. R. James*, ed. Richard Dalby and Rosemary Pardoe (1987).
23. Arthur Conan Doyle, *The Valley of Fear* (1914), ch. 6. A good deal of crime fiction makes use of the preternatural as an element in the process of mystification. Conan Doyle's own *The Hound of the Baskervilles* (1902) still has power to raise a shiver, while Margery Allingham's *Look to the Lady* (1931) contains an overtly supernaturalist element in a traditional Gothic setting; in this case the secret in the locked room is no anticlimax when revealed.

Chapter Three: Watchers on the Threshold

1. Alexander Gilchrist, *The Life of William Blake* (1863), ch. 13.
2. Horatio F. Brown, *John Addington Symonds: A Biography* (1895), 7–8.
3. James Webb, *The Flight from Reason* (1971), 101.
4. Marie M. Tatar, *Spellbound: Essays on Mesmerism and Literature* (Princeton, NJ, 1978), 47–8.
5. Ibid. 49.
6. For an account of these hauntings, see Sacheverell Sitwell, *Poltergeists: An Introduction and Examination Followed by Chosen Incidents* (1940), 189–214.
7. Where Mrs Poyntz is concerned, Lytton achieves a magnificent absurdity: 'I sprang to my feet, with difficulty suppressing my rage; and, remembering it was a woman who spoke to me, "Farewell, madam," said I, through my grinded teeth. "Were you, indeed, the Personation of the World, whose mean notions you mouth so calmly, I could not disdain you more." I turned to the door, and left her still standing erect and menacing, the hard sneer on her resolute lip, the red glitter in her remorseless eye' (ch. 57). 'Ouida' herself could not have bettered that.
8. *Supernatural Horror in Literature* (New York, 1973), 82.

9. *Là-Bas*, ch. 18. It is worth noting in this connection that George Eliot published a story about a clairvoyant in the July 1859 issue of *Blackwood's Magazine*. 'The Lifted Veil' is unsuccessful as a tale of the paranormal or occult, asking to be read more as a parabolic psychological analysis of the author's own anxieties concerning her powers as a novelist and her moral responsibilities as an artist.

10. *The Watcher by the Threshold and Other Tales* (1902). Buchan was more interested in history than in the occult. His account of Satanism in the Scottish borders, *Witch Wood* (1927), is focused on the 17th-cent. believers rather than on the nature of their beliefs.

11. In their stress upon physical loathsomeness Machen's early tales anticipate the writings of the New England master of the genre, H. P. Lovecraft (1890–1937) (whose Cthulu mythology bears certain resemblances to Machen's 'Little People'), but they are far more elegantly written.

12. Prologue, *The Three Imposters*.

13. The book was started in the 1890s and was to have been called 'The Garden of Avallaunius'. Its substituted title is misleading, being reminiscent of the soporific books of popular philosophy by A. C. Benson (*The House of Quiet* (1904), *The Thread of Gold* (1905), *Beside Still Waters* (1907), and so on) which were being published at the time. (A not dissimilar misfortune had already befallen George MacDonald, whose masterwork *Lilith* was immediately preceded by *The Soul of Lilith* (1894) by Marie Corelli, with which pretentious saga it was to be embarrassingly compared. Corelli's own supernaturalist fictions (*The Sorrows of Satan* (1895), *Ziska, the Problem of a Wicked Soul* (1897), etc.) enjoyed enormous popularity at the time.

14. 'The White People' was included, with *The Great God Pan*, *The Three Imposters*, and three other stories, in *The House of Souls* (1906). One can only hope that no admirer of A. C. Benson's essays was mistakenly given this book as a present.

15. William Hope Hodgson (1877–1918) specialized in horror stories of the sea. His fantasy novels *The House on the Borderland* (1908) and *The Night Land* (1912) reveal a powerful but undiscriminating imagination. Discrimination, however, is of the essence of *Carnaki the Ghost Finder*, in which some of the manifestations turn out to be genuine, others to be materialistically explicable.

16. The spelling was also that of A. E. Waite, the Christian occult scholar to whose Rosicrucian Order Machen for a while belonged. The spelling is also found in novels by two other members, Evelyn Underhill's *The Column of Dust* (1909) and Charles Williams's *War in Heaven* (1930).

17. He was a member of A. E. Waite's Rosicrucian Order. See R. A. Gilbert, *The Golden Dawn: Twilight of the Magicians* (Wellingborough, 1983). The author's contention that 'The only certain case of the ideas and practices of the Golden Dawn moulding the whole work of an author is that of Algernon Blackwood' (p. 88) needs to be qualified

in the light of the latter's own comments in ch. 5 of *Episodes Before Thirty*.

18. *Golden Dawn*, 82.
19. *Supernatural Horror*, 96.
20. Quoted in Bernard Sellin, *The Life and Work of David Lindsay*, tr. Kenneth Gunnell (Cambridge, 1961), 31.
21. J. B. Pick, Introduction to *The Haunted Woman*, Canongate Classics edn. (Edinburgh, 1987), p. xii.
22. Lindsay's subsequent novels nowhere surpass the power of their two predecessors. *Sphinx* (1923) deals as much with the paranormal as with the supernatural: its story of a man's experiments in recording his own dreams cinematically is novelettish and rather trite: it affords no real sense of the mysterium. Two novels were left unpublished at Lindsay's death. The unfinished *The Witch* concerns the power of music; in *The Violet Apple* (1973) a pip from the Tree of Life is discovered as the result of a seeming accident, and grows into a tree. Here the mythological is more happily integrated into a story of human love and personal affinities.
23. The pseudonyms were 'Miss Morison' and 'Miss Lamont'. The true names were given in the 4th edn. For a critical account of the affair, see Lucille Iremonger, *The Ghosts of Versailles* (1957).

Chapter Four: An Insinuation of Doctrine

1. Rolland Hein, *The Harmony Within: The Spiritual Vision of George MacDonald* (Grand Rapids, Mich., 1982), 113.
2. C. N. Manlove, *Modern Fantasy* (Cambridge, 1975), 71.
3. W. H. Auden, *Forewords and Afterwords* (1973), 270.
4. William Raeper, *George MacDonald* (Tring, 1987), 209.
5. *The Victorian Age in Literature* (1913), 145.
6. *The Descent of the Dove: A Short History of the Holy Spirit in the Church* (1939), 219. 'Almighty God' was a term much in use among middle-of-the-road Anglicans, as distinct from the Anglo-Catholic preference for 'Our Lord' and the Evangelicals' use of 'Christ'. It reflects, however unconsciously, a concern with the stabilizing, hierarchical character of the Established Church. Williams is ironically aware of this.
7. Evelyn Underhill (1875–1941) was an influential lay theologian, devotional writer, and authority on mysticism. Her three novels (see Bibliography), like those of Charles Williams, reveal the interplay of occultism with a Christian interpretation of reality.
8. 'Religion appeared, poor little talkative Christianity, and she knew that all its divine words from "Let there be Light" to "It is finished" only amounted to "boum"' (*A Passage to India*, ch. 15).

Chapter Five: Twilight Territories

1. *Letters of Max Beerbohm 1892–1956*, ed. Rupert Hart-Davis (1988), 94.

2. *Life's Handicap: Being Stories of Mine Own People* (1891).
3. Ibid.
4. *Wee Willie Winkie and Other Stories* (1888).
5. Ibid.
6. Ibid.
7. *Life's Handicap.*
8. *Debits and Credits* (1926).
9. Ibid.
10. *Many Inventions* (1893).
11. *Traffics and Discoveries* (1904).
12. *Actions and Reactions* (1909).
13. Ibid.
14. *Something of Myself* (1937), 185–6.
15. 'Let England be Sir Philip Sidney, Shakespere, Milton, Bacon, Harrington, Swift, Wordsworth; and never let the names of Darwin, Johnson, Hume, *fur* it over. If these, too, must be England, let them be another England; or, rather, let the first be old England, the spiritual, Platonic old England, and the second, with Locke at the head of the philosophers and Pope (at the head) of the poets . . . be the representatives of Commercial Great Britain', *Anima Poetae*, ed. Ernest Hartley Coleridge (1895), 151. The dichotomy was to strike a resounding chord in the imaginations of many 20th-cent. novelists and poets, including Chesterton, Buchan, Charles Williams, and Geoffrey Hill. (It has also provided an insidious lure to nostalgic sentimentalists.)
16. *Puck of Pook's Hill* (1906).
17. *Rewards and Fairies* (1910).
18. *Traffics and Discoveries.*
19. Richard Middleton, *The Ghost Ship and Other Stories* (1912).
20. *Debits and Credits.*
21. *A Diversity of Creatures* (1917).
22. de la Mare's second full-length prose fiction, *The Three Mulla Mulgars* (1910), renamed *The Three Royal Monkeys*, is a book for children in which a genuine sense of the supernatural, full of mystery and awe, is conveyed within a work of fantasy—an unusual achievement.
23. 'The Listeners', *The Listeners and Other Poems* (1912).
24. 'The Ghost', *Motley and Other Poems* (1912).
25. *Ding Dong Bell* (1924).
26. *The Wind Blows Over* (1936).
27. *On the Edge: Short Stories* (1930).
28. *A Beginning and Other Stories* (1955).
29. *On the Edge.*
30. Ibid.
31. 'Seaton's Aunt', *The Riddle and Other Stories* (1922).
32. *On the Edge.*
33. *A Beginning.*

34. *On the Edge.*
35. *Broomsticks and Other Tales* (1925).
36. *The Riddle.*
37. *The Wind Blows Over.*
38. *On the Edge.*
39. *The Riddle.*
40. *The Connoisseur and Other Stories* (1926).
41. *The Wind Blows Over.*
42. *The Connoisseur.*
43. *The Riddle.*
44. Ibid.
45. *The Connoisseur.*
46. Graham Greene, *The Lost Childhood* (1951), 82.
47. *Ding Dong Bell.*
48. *The Riddle.*
49. 'Dejection: An Ode'.
50. *The Riddle.*
51. *A Beginning.*
52. *Broomsticks.*
53. *On the Edge.*
54. 'Walter de la Mare', *Living Writers*, ed. Gilbert Phelps (1947), 117.

Chapter Six: Numinous Landscapes

1. See Kenneth Clark, *Landscape into Art* (new edn. 1976), ch. 5.
2. *The Seven Lamps of Architecture* (1849), 2nd preface.
3. *The Task*, book 1.
4. Florence Emily Hardy, *The Life of Thomas Hardy* (1962 edn.), 370.
5. Ibid., 309.
6. *The Story of My Heart* (1883), ch. 3.
7. Ibid., ch. 6.
8. *The Celestial Omnibus* (1911). The story was written in 1904.
9. *Reginald in Russia* (1910).
10. *The Chronicles of Clovis* (1911).
11. See e.g. ch. 12, 'The Great Passing' in *The Boy in the Bush*, Lawrence's rewriting of the novel by M. L. Skinner.
12. *Landscape into Art*, 230.
13. Raymond Williams, *The Country and the City* (1973), 254. Williams was himself the author of a novel with a powerful rendering of timelessness in a particular place, *People of the Black Mountain* (1989).
14. *The Old Road from Spain* (The World's Classics edn. 1932), p. xii.
15. *The Saturnian Quest: A Chart of the Prose Works of John Cowper Powys* (1964), 28–9.
16. A sampling of fields in the parish of North Cadbury in Somerset yields such names as Gillon's Lawn, Peace, Clare Hill, Dodinells, God's Hill, Purgatory, and Herne. (I am indebted for this information to Mr Angus Robson.)

17. John A. Brebner, *The Demon Within: A Study of John Cowper Powys's Novels* (1973), 49.
18. *A Glastonbury Romance* (3rd edn. 1955), p. xv.
19. H. W. Fawkner, *The Ecstatic World of John Cowper Powys* (1986), 216.
20. *Maiden Castle* was first published in truncated form. The full text was published in 1990 by the Univ. of Wales Press (Cardiff).
21. Like *Maiden Castle*, but even more drastically, the text of *Porius* was mutilated before publication. A complete version was published by Colgate Univ. Press (Hamilton, NY) in 1994.
22. *Autobiography* (1934), 641–2.
23. Fawkner, *Ecstatic World*, 23–4.
24. Denis Lane, 'The Elemental Image in *Wolf Solent*', in Denis Lane (ed.), *In the Spirit of Powys* (1990), 58.
25. *Anima Poetae*, 136.

Chapter Seven: The Enemy Within

1. *The White Wand and Other Stories* (1954).
2. John Galt (1779–1839) is another instance of a naturalistic novelist who on occasion was attracted towards the supernatural—see e.g. *The Spae-Wife* (1823) and *The Omen* (1825).
3. *The English Novel* (1954), 131.
4. *The Merry Men and Other Tales and Fables* (1886).
5. Leon Edel, *Henry James: The Untried Years 1843–1870* (1953), 31.
6. *The Complete Tales of Henry James*, ed. Leon Edel (1962–4), i.
7. Ibid. iii.
8. Ibid. xi.
9. Ibid. ix.
10. Ibid. iv.
11. Ibid. ix. The renaming occurred in the New York edn. of James's novels and tales (1908).
12. Ibid. x.
13. Ibid. xii.
14. James has his own way of testing his readers. 'Isn't it a part of what I call the beauty that this concomitant, this watchful and critical, living in his "own" self inevitably grows and grows from a certain moment on?—and isn't it for instance quite magnificent that one sees this growth of it as inevitably promoted more and more by his sense of what I have noted as the malaise on the part of the others? Don't I see his divination and perception of *that* so affect and act upon him that little by little he begins to love more, to live most, and most uneasily, in what I refer to as his own, his prior self, and less, uneasily less, in his borrowed, his adventurous, that of his tremendous speculation, so to speak—rather than the other way round as has been the case at first', *The Sense of the Past* (1917), 294–5. Mystification becomes awesome enough in this passage as almost to amount to mystery.
15. *Complete Tales*, x.

16. 'When the unclean spirit is gone out of a man, he walketh through dry places, seeking rest and finding none. Then he saith, I will return into my house from whence I came out; and when he is come, he findeth it empty, swept and garnished. Then goeth he, and taketh with himself seven other spirits more wicked than himself, and they enter in and dwell there: and the last state of that man is worse than the first', Matthew 12: 43–5.

17. *The Art of the Novel* (1947), 176.

18. It is noteworthy that all the characters surrounding the children are employees: the threat posed by the servant class to the absentee landlord adds a note of collusive social rebellion to the menace of the tale. The social position of governesses was always ambiguous, witness *Jane Eyre* and *Agnes Grey*.

19. *Complete Tales*, xi.

20. Ibid. ix.

21. 'Henry James: The Religious Aspect', *The Lost Childhood*, 55.

22. Elizabeth Bowen, *Afterthought: Pieces about Writing* (1962), 102. Cf. William Empson: 'the ghost of spiritualism . . . is too strong. For if the real supernatural is as silly as that, and a ghost story depending on the popular beliefs of the time would have to spend all its time psychoanalysing the ghost, where can one flee but to the terrors of the soul, and to the reader psychoanalysing the author?' *Empson in Granta 1927–1929* (1993).

23. Cynthia Asquith (ed.), *The Second Ghost Book* (1952). Repr. in *Collected Stories* (1980).

24. *The Cat Jumps* (1934).

25. 'Mysterious Kor', *The Demon Lover* (1945).

26. *The Tatler and Bystander*, 226/2945. Bowen likens the novel to Le Fanu's *Uncle Silas* (a book she much admired) and suggests the contemporary interest Paul's work aroused by remarking that 'many a critic still harks back to *Camilla*'.

27. Phyllis Paul died on 30 Aug. 1973, in Hastings, as a result of being struck by a motor cycle while crossing the road. The account at the inquest suggests that she was not known locally as a writer, being only identified by the Cash name tag on her handkerchief. A neighbour commented that 'Miss Paul kept herself to herself. When she walked she had a habit of looking quickly to one side and then the other, and then she would look down again.' A witness to the accident was more graphic still, remarking that what he saw was 'an old lady going across the road like a sheet of newspaper'. The phrase might have been coined by Paul herself (see *Hastings Observer*, 8 and 15 Sept. 1973).

28. Luke 16: 31.

29. 1 Peter 5: 8.

30. Lucy Snowe is the passionate, solitary narrator of Charlotte Brontë's *Villette* (1853). Paul's critical view of ill-brought-up children is shared by Anne Brontë and Jane Austen.

Chapter Eight: God-Games

1. *Lord Arthur Saville's Crime and Other Stories* (1891).
2. *Cold Comfort Farm* (1932) by Stella Gibbons was a highly popular satire on certain rural writers of the time. It made fun of, among others, D. H. Lawrence, John Cowper and T. F. Powys, and Mary Webb.
3. In his comic fantasy novel *Mr Pye* (1953) Peake does introduce the supernatural, albeit in light-hearted form. According to angelic or diabolic inclinations, his protagonist sprouts wings or horns.
4. *The White Paternoster and Other Stories* (1930).
5. A number of Baker's novels deal with the preternatural, but he sounds the note of fantasy to most sinister effect in *The Birds* (1936), which both in theme and title anticipates the tale by Daphne du Maurier, filmed by Alfred Hitchcock in 1963. It combines social satire with apocalyptic fable. These birds are harpies, the avenging powers of humankind's rejected self; both in ideas and in the forcefulness of its climax, when the birds invade St Paul's Cathedral during a service of thanksgiving for their departure, the book recalls Charles Williams's *The Place of the Lion*.
6. Introduction, *The Magus* (revised edn. 1977).
7. Cf. Isak Dinesen, 'The Immortal Story', *Anecdotes of Destiny* (1958).
8. I am indebted for this observation to Canon J. W. Lee.
9. Elizabeth Dipple, *Iris Murdoch: Work for the Spirit* (1982), 86.
10. Cf. John 20: 17, the risen Christ's words to Mary Magdalene.
11. 'Miss Pinkerton's Apocalypse', *The Go-Away Bird* (1958).
12. *The Bachelors* (1960), ch. 6.
13. This turning of characters into the kind of monsters that George Orwell notes in the work of Dickens (*Collected Essays* (1961), 81–2) occurs in Spark's fiction more in the shape of comical names that 'place' people: e.g. Dawn Wagstaffe, Grey Mauser, Tempest Sidebottome, Huy Throvis-Mew, and so on. She rivals Ronald Firbank in this respect.
14. Frank Kermode, 'The House of Fiction: Interviews with Seven English Novelists' (*Partisan Review*, 30 (Spring 1963), 79–82). Repr. in Malcolm Bradbury (ed.), *Contemporary Writers on Modern Fiction* (1977).
15. *The Mandelbaum Gate* (1965), ch. 6.
16. *The Letters of Evelyn Waugh* (1980), ed. Mark Amory, 477.

Chapter Nine: Reversions to Type

1. This may be because, as Dr Paul Hartle points out to me, the line constitutes an iambic hexameter.
2. Forrest Reid, *Tom Barber* (New York, 1955), introduction by E. M. Forster, p. 9.
3. *Apostate* (1926), ch. 19.
4. *Uncle Stephen* (1931), ch. 8.
5. Ibid., ch. 15.

6. Neil Philip, *A Fine Anger* (1981), 46.
7. *The Owl Service* re-enacts the legend of Blodeuwedd, as recounted in 'Math, Son of Mathonwy', Branch Four of the *Mabinogion*.
8. Philip, *Fine Anger*, 129–30.
9. Gunn's work is full of a sense of the numinous qualities of particular places, and one might also cite *The Well at the World's End* (1951) and *The Other Landscape* (1954) as instances of the romanticism of his later fiction.

Chapter Ten: The Earthing of the Supernatural

1. Virginia Woolf, 'The Leaning Tower', *The Moment and Other Essays* (1947), 109.
2. *Metaphysics as a Guide to Morals* (1992), 235.
3. Mrs Oliphant, 'The Open Door', *Stories of the Seen and Unseen* (1902). The two preceding manifestations occur in William Fryer Harvey, 'The Clock', *The Beast with Five Fingers* (1928), and in Oliver Onions, 'The Beckoning Fair One', *Widdershins* (1911).
4. *The Diary of John Cowper Powys 1931* (1990), 305.
5. '. . . May God us keep From Single vision & Newton's sleep!' from letter to Thomas Butts, 22 Nov. 1802.
6. C. G. Jung, *Memories, Dreams, Reflections* (1963), tr. Richard and Clara Winston, ch. 12, ii.
7. *Real Presences: Is There Anything In What We Say?* (1989), 215.
8. 'The Supernatural in Fiction', *Granite and Rainbow* (1958), 62.

Epilogue: An Abolishing of Idols

1. *Agnes Grey* (ch. 1).
2. 'The Story', *Aspects of the Novel* (1927), ch. 2.
3. A good deal of contemporary literary criticism reflects the working conditions of university teachers, whose professional lives revolve around seminars and academic conferences in an atmosphere of economic competition that requires the publication of books and articles for purposes of tenure and promotion. Such an atmosphere is scarcely conducive to the consideration of matters imaginative and transcendental.
4. This state of being, or providential absolute, may still be legitimately designated 'God', since that is a neutral word, free of linguistic associations of a contemporary kind. When harnessed by human prejudices and ideologies and desires, however, it automatically turns into an idol.
5. Charles Lock, 'Against Being: An Introduction to the Thought of Jean-Luc Marion', *St Vladimir's Theological Quarterly*, 37/4 (1993), 371.
6. *Chance* (1913), part ii, ch. 4. By contrast, in *The Shadow Line* (1917) Conrad provides a convincing and at the same time semi-humorous portrayal of preternatural dread.
7. Charles Williams, *The House of the Octopus* (1945), Act ii: *Collected Plays* (Oxford, 1963), 287.

BIBLIOGRAPHY

Part One contains a selective list of novels which treat of supernaturalist themes or which contain supernaturalist elements. Part Two contains a selective list of collections either wholly or partially consisting of supernaturalist tales. Part Three contains a selective list of works of general interest in connection with the writings listed and discussed above. Unless otherwise stated, the place of publication is London.

Part One: Novels

Ackroyd, Peter, *Hawksmoor* (1985); *Chatterton* (1987); *First Light* (1989); *English Music* (1992); *The House of Doctor Dee* (1993).

Ainsworth, William Harrison, *Rookwood* (1834); *Windsor Castle* (1843); *The Lancashire Witches* (1848); *Auriol: or The Elixir of Life* (1865).

Allingham, Margery, *Look to the Lady* (1931).

Amis, Kingsley, *The Green Man* (1969).

Baker, Frank, *The Birds* (1936); *Miss Hargreaves* (1940); *The Downs So Free* (1948); *Talk of the Devil* (1956).

Bennett, Arnold, *The Ghost* (1907); *The Glimpse* (1909).

Benson, E. F., *The Inheritor* (1930).

Benson, R. H., *Lord of the World* (1908); *The Necromancers* (1909); *The Dawn of All* (1911); *Initiation* (1914).

Blackwood, Algernon, *Jimbo* (1909); *The Human Chord* (1910); *The Centaur* (1911); *The Promise of Air* (1918); *The Education of Uncle Paul* (1920).

Bowen, Elizabeth, *A World of Love* (1955).

Bridge, Ann, *And Then You Came* (1948).

Briggs, K. M., *Hobberdy Dick* (1955); *Kate Crackernuts* (1963).

Brontë, Charlotte, *Jane Eyre* (1847).

Brontë, Emily, *Wuthering Heights* (1847).

Buchan, John, *The Dancing Floor* (1926); *Witch Wood* (1927); *The Gap in the Curtain* (1932).

Bulwer-Lytton, Sir Edward, *Godolphin* (1833); *Zanoni* (1842); *A Strange Story* (1862).

Chesterton, G. K., *The Man Who Was Thursday* (1908).

Collins, Wilkie, *Armadale* (1864); *The Two Destinies* (1876); *The Haunted Hotel* (1877).

Cooper, Susan, *Over Sea, Under Stone* (1965); *The Dark is Rising* (1966); *Greenwitch* (1974); *The Grey King* (1975); *Silver on the Tree* (1977).

de la Mare, Walter, *The Return* (1910).

Dickens, Charles, *Bleak House* (1851).

du Maurier, Daphne, *The House on the Strand* (1969).

du Maurier, George, *Peter Ibbetson* (1892).

Dunsany, Lord, *The Curse of the Wise Woman* (1933).

Falkner, J. Meade, *The Lost Stradivarius* (1895).

Ferguson, Rachel, *A Harp in Lowndes Square* (1936).

Ford, Ford Madox, *Ladies Whose Bright Eyes* (1911).

Fortune, Dion, *The Goat-Foot God* (1936).

Fowles, John, *The Magus* (1966).

Galt, John, *The Spae-Wife* (1823); *The Omen* (1825).

Garner, Alan, *The Weirdstone of Brisingamen* (1960); *The Moon of Gomrath* (1963); *Elidor* (1965); *The Owl Service* (1967); *Red Shift* (1973).

Garnett, David, *Lady into Fox* (1922).

Godwin, William, *St Leon* (1800).

Golding, William, *The Spire* (1964); *The Paper Men* (1984).

Gordon, John, *The House on the Brink* (1970).

Gunn, Neil M., *Second Sight* (1940); *The Silver Bough* (1947); *The Well at the World's End* (1951); *The Other Landscape* (1954).

Hartley, L. P., *Eustace and Hilda* (1947).

Hill, Susan, *The Woman in Black* (1983); *The Mist in the Mirror* (1992).

Hogg, James, *The Private Memoirs and Confessions of a Justified Sinner* (1827).

Holme, Constance, *The Old Road from Spain* (1915); *He-Who-Came?* (1930).

Irwin, Margaret, *Still She Wished for Company* (1924).

Laski, Marghanita, *The Victorian Chaise-Longue* (1953).

Lewis, C. S., *That Hideous Strength* (1945); *Till We Have Faces* (1956).

Lewis, Matthew Gregory, *The Monk* (1796).

Lindsay, David, *The Haunted Woman* (1922); *Sphinx* (1923); *Devil's Tor* (1932); *The Violet Apple* (1973); *The Witch* (1973).

Macaulay, Rose, *The Secret River* (1909).

MacDonald, George, *The Portent* (1860); *Adela Cathcart* (1864); *At the Back of the North Wind* (1871); *The Princess and the Goblin* (1872); *The Wise Woman* (1875); *The Princess and Curdie* (1883); *Lilith* (1895).

Machen, Arthur, *The Hill of Dreams* (1907); *The Terror* (1917); *The Secret Glory* (1922); *The Green Round* (1933).

Manning, Rosemary, *Open the Door* (1983).

Marryat, Frederick, *The Phantom Ship* (1839).

Marsh, Richard, *The Beetle* (1897).

Maturin, Charles, *Melmoth the Wanderer* (1820).

Murdoch, Iris, *The Nice and the Good* (1968); *The Sea, the Sea* (1978); *Nuns and Soldiers* (1980); *The Good Apprentice* (1985); *The Message to the Planet* (1989); *The Green Knight* (1993).

Oliphant, Margaret, *A Beleagured City* (1880); *A Little Pilgrim in the Unseen* (1882).

Olivier, Edith, *The Love-Child* (1927).

Pater, Walter, *Marius the Epicurean* (1885).

Paul, Phyllis, *Camilla* (1949); *The Lion of Cooling Bay* (1953); *Rox Hall Illuminated* (1955); *Twice Lost* (1960); *A Little Treachery* (1962).

Peake, Mervyn, *Mr Pye* (1953).

Pearce, Philippa, *Tom's Midnight Garden* (1958).

Powys, John Cowper, *Ducdame* (1925); *A Glastonbury Romance* (1933); *Maiden Castle* (1936); *Morwyn* (1937); *Owen Glendower* (1941); *Porius* (1951).

Powys, T. F., *Mockery Gap* (1925); *Mr Weston's Good Wine* (1927); *Unclay* (1931).

Radcliffe, Ann, *Gaston de Blondeville* (1826).

Reid, Forrest, *The Bracknels* (1911); *Uncle Stephen* (1931).

Scott, Sir Walter, *Waverley* (1814); *Guy Mannering* (1815); *The Antiquary* (1816); *The Black Dwarf* (1816); *The Bride of Lammermoor* (1819); *A Legend of Montrose* (1819); *The Monastery* (1820); *Redgauntlet* (1824); *The Betrothed* (1825).

Shelley, Mary, *Frankenstein* (1815).

Sitwell, Osbert, *The Man Who Lost Himself* (1930); *A Place of One's Own* (1941).

Somerville, E., and Ross, M., *The Silver Fox* (1898).

Spark, Muriel, *The Comforters* (1957); *Memento Mori* (1959); *The Ballad of Peckham Rye* (1960); *The Hothouse by the East River* (1973).

Stevenson, Robert Louis, *The Strange Case of Dr Jekyll and Mr Hyde* (1886).

Stoker, Bram, *Dracula* (1897); *The Lair of the White Worm* (1911).

Underhill, Evelyn, *The Grey World* (1904); *The Lost Word* (1907); *The Column of Dust* (1909).

Walpole, Horace, *The Castle of Otranto* (1764).

Walpole, Hugh, *Maradick at Forty* (1910); *The Prelude to Adventure* (1912); *The Fortress* (1932); *The Killer and the Slain* (1942).

Warner, Sylvia Townsend, *Lolly Willowes* (1926).

Waugh, Evelyn, *Helena* (1950); *The Ordeal of Gilbert Pinfold* (1957).

Webb, Mary, *The Golden Arrow* (1915); *Gone to Earth* (1917); *Precious Bane* (1924).

Wilde, Oscar, *The Picture of Dorian Gray* (1891).

Williams, Charles, *War in Heaven* (1930); *Many Dimensions* (1931); *The Place of the Lion* (1931); *The Greater Trumps* (1932); *Shadows of Ecstasy* (1933); *Descent into Hell* (1937); *All Hallows' Eve* (1945).

Wood, Mrs Henry, *The Shadow of Ashlydyat* (1863).

Young, Francis Brett, *Undergrowth* (with E. Brett Young, 1913); *The Crescent Moon* (1918); *Cold Harbour* (1924).

Part Two: Collections

Aickman, Robert, *We Are For The Dark* (1951); *Dark Entries* (1964); *Cold Hand in Mine* (1975).

Benson, A. C., *The Hill of Trouble* (1903); *The Isles of Sunset* (1904).

Benson, E. F., *The Room in the Tower* (1912); *Visible and Invisible* (1923); *Spook Stories* (1928); *More Spook Stories* (1934).

Benson, R. H., *The Light Invisible* (1903); *A Mirror of Shalott* (1907).

Blackwood, Algernon, *The Empty House* (1906); *The Listener* (1907); *John Silence, Physician Extraordinary* (1908); *Strange Stories* (1929).

Bowen, Elizabeth, *The Cat Jumps* (1934); *The Demon Lover* (1945).

Broughton, Rhoda, *Twilight Stories* (1879).

Buchan, John, *Grey Weather* (1899); *The Watcher by the Threshold* (1902); *The Moon Endureth* (1912); *The Runagates Club* (1928).

Collins, Wilkie, *After Dark* (1856); *The Queen of Hearts* (1859).

Coppard, A. E., *Fearful Pleasures* (1951).

Crawford, F. Marion, *Uncanny Tales* (1911).

Crowe, Catherine, *The Night Side of Nature* (1848).

de la Mare, Walter, *The Riddle* (1922); *Ding Dong Bell* (1924); *Broomsticks* (1925); *The Connoisseur* (1926); *On the Edge* (1930); *The Wind Blows Over* (1936); *A Beginning* (1955); *Eight Tales* (Sauk City, Wis., 1971).

Dickens, Charles, *Christmas Books* (1845–7); *Christmas Stories* (1850–67).

Doyle, Sir Arthur Conan, *Dreamland and Ghostland* (1885).

du Maurier, Daphne, *Not After Midnight* (1971).

Forster, E. M., *The Celestial Omnibus* (1911); *The Eternal Moment* (1928).

Gaskell, Elizabeth, *Lizzie Leigh and Other Tales* (1855); *Round the Sofa* (1859); *The Grey Woman and Other Tales* (1865).

Hardy, Thomas, *Wessex Tales* (1888); *A Changed Man* (1913).

Hartley, L. P., *The Travelling Grave* (1951).

Harvey, W. F., *Midnight Tales* (1946).

Hodgson, William Hope, *Carnacki, The Ghost-Finder* (1913).

Howard, Elizabeth Jane, *We Are For The Dark* (1951); *Mr Wrong* (1975).

Irwin, Margaret, *Madame Fears the Dark* (1935).

James, Henry, *The Ghostly Tales of Henry James* (1948).

James, Montague Rhodes, *Collected Ghost Stories* (1931).

Kipling, Rudyard, *Wee Willie Winkie* (1888); *Life's Handicap* (1891); *Traffics and Discoveries* (1904); *Puck of Pook's Hill* (1906); *Actions and Reactions* (1909); *Rewards and Fairies* (1910); *Debits and Credits* (1926).

Lawrence, D. H., *The Lovely Lady* (1933).

Lee, Vernon, *Hauntings* (1890).

Le Fanu, J. Sheridan, *Ghost Stories and Tales of Mystery* (Dublin, 1851); *In a Glass Darkly* (1872); *The Purcell Papers* (1880); *Madam Crowl's Ghost* (ed. M. R. James, 1923).

Machen, Arthur, *The House of Souls* (1906); *Holy Terrors* (Harmondsworth, 1947).

Malden, R. H., *Nine Ghosts* (1943).

Middleton, Richard, *The Ghost Ship* (1912).

Munby, A. N. L., *The Alabaster Hand* (1949).

Nesbit, Edith, *Grim Tales* (1893); *Fear* (1910).

Oliphant, Margaret, *The Land of Darkness* (1886); *Stories of the Seen and Unseen* (1889).

Onions, Oliver, *Widdershins* (1911); *Ghosts in Daylight* (1924).

Powys, T. F., *The Left Leg* (1923), *The White Paternoster* (1930).

Riddell, Charlotte, *Weird Stories* (1884).

Rolt, L. T. C., *Sleep No More* (1948).

Saki, *Reginald in Russia* (1910); *The Chronicles of Clovis* (1911); *Beasts and Super-Beasts* (1914); *The Toys of Peace* (1923).

Sinclair, May, *Uncanny Stories* (1923).
Stevenson, Robert Louis, *The Merry Men* (1886).
Swain, E. G., *The Stoneground Ghost Tales* (Cambridge, 1912).
Wakefield, H. Russell, *They Return at Evening* (1928); *Old Man's Beard* (1929).
Walpole, Hugh, *All Souls' Night* (1933).
White, T. H., *The Maharajah* (1981).
Wilde, Oscar, *Lord Arthur Savile's Crime* (1887).
Woodforde, Christopher, *A Pad in the Straw* (1952).
Yeats, W. B., *The Celtic Twilight* (1893).

Part Three: Works of General Interest

Abrams, M. H., *Natural Supernaturalism: Tradition and Revolution in Romantic Literature* (Oxford, 1971).
Ashley, Mike, *Who's Who in Horror and Fantasy Fiction* (1977).
Baine, Rodney M., *Daniel Defoe and the Supernatural* (Athens, Ga., 1968).
Birkhead, Edith, *The Tale of Terror* (1921).
Briggs, Julia, *Night Visitors: The Rise and Fall of the English Ghost Story* (1977).
Ellis, S. M., *Wilkie Collins, Le Fanu and Others* (1931).
Freud, Sigmund, *The Uncanny* (Collected Papers, 4; 1957).
Jackson, Rosemary, *Fantasy: The Literature of Subversion* (1981).
James, M. R., Introduction to *Ghosts and Marvels*, ed. V. H. Collins (Oxford, 1924).
Kermode, Frank, *The Genesis of Secrecy: On the Interpretation of Narrative* (1979).
Leatherdale, Clive, *The Origin of Dracula* (1987).
Lovecraft, H. P., *Supernatural Horror in Literature* (New York, 1945).
Machen, Arthur, *Hieroglyphics* (1902).
Manlove, C. N., *Modern Fantasy: Five Studies* (1973).
Mehotra, K. M., *Horace Walpole and the English Novel* (Oxford, 1934).
Parsons, Coleman C., *Witchcraft and Demonology in Scott's Fiction* (1964).
Penzoldt, Peter, *The Supernatural in Fiction* (1952).
Prickett, Stephen, *Victorian Fantasy* (Hassocks, 1979).
Sage, Victor, *Horror Fiction in the Protestant Tradition* (1988).
Saurat, Denis, *Literature and the Occult Tradition* (1930).
Scarborough, Dorothy L., *The Supernatural in Modern English Fiction* (1917).
Steiner, George, *Real Presences: Is There Anything in What We Say?* (1989).
Sullivan, Jack, *Elegant Nightmares: The English Ghost Story from Le Fanu to Blackwood* (Ohio, 1978).
Summers, Montague, Introduction to *The Supernatural Omnibus* (1936); *The Gothic Quest* (1938).
Webb, James, *The Age of the Irrational: The Flight from Reason* (1971).
Woolf, Virginia, *Granite and Rainbow* (1958).

INDEX